SpringerBriefs in Earth Sciences

For further volumes:
http://www.springer.com/series/8897

Andrew Y. Glikson

Evolution of the Atmosphere, Fire and the Anthropocene Climate Event Horizon

With a forward by Professor
H. J. Schellnhuber
Director, Potsdam Institute of Climate
Impact Research (PIK)

 Springer

Andrew Y. Glikson
School of Archaeology and Anthropology
Australian National University
Canberra, ACT
Australia

ISSN 2191-5369 ISSN 2191-5377 (electronic)
ISBN 978-94-007-7331-8 ISBN 978-94-007-7332-5 (eBook)
DOI 10.1007/978-94-007-7332-5
Springer Dordrecht Heidelberg New York London

Library of Congress Control Number: 2013944539

The facts and opinions expressed in this work are those of the author and not necessarily of the publisher.

© The Author(s) 2014
This work is subject to copyright. All rights are reserved by the Publisher, whether the whole or part of the material is concerned, specifically the rights of translation, reprinting, reuse of illustrations, recitation, broadcasting, reproduction on microfilms or in any other physical way, and transmission or information storage and retrieval, electronic adaptation, computer software, or by similar or dissimilar methodology now known or hereafter developed. Exempted from this legal reservation are brief excerpts in connection with reviews or scholarly analysis or material supplied specifically for the purpose of being entered and executed on a computer system, for exclusive use by the purchaser of the work. Duplication of this publication or parts thereof is permitted only under the provisions of the Copyright Law of the Publisher's location, in its current version, and permission for use must always be obtained from Springer. Permissions for use may be obtained through RightsLink at the Copyright Clearance Center. Violations are liable to prosecution under the respective Copyright Law.
The use of general descriptive names, registered names, trademarks, service marks, etc. in this publication does not imply, even in the absence of a specific statement, that such names are exempt from the relevant protective laws and regulations and therefore free for general use.
While the advice and information in this book are believed to be true and accurate at the date of publication, neither the authors nor the editors nor the publisher can accept any legal responsibility for any errors or omissions that may be made. The publisher makes no warranty, express or implied, with respect to the material contained herein.

Printed on acid-free paper

Springer is part of Springer Science+Business Media (www.springer.com)

In honor of James Hansen and David Attenborough

Foreword

When I read Andrew Glikson's new book the famous Russell–Einstein Manifesto, published in 1955 in the midst of the Cold War, came to my mind. In this manifesto, Bertrand Russell, Albert Einstein and the other distinguished signatories point to the massive dangers of nuclear weapons and call for peaceful solutions, while stating: "Many warnings have been uttered by eminent men of science … We have not yet found that the views of experts on this question depend in any degree upon their politics or prejudices. They depend only … upon the extent of the particular expert's knowledge. We have found that the men who know most are the most gloomy". Indeed, Andrew Glikson shares an enormous wealth of knowledge with us, putting anthropogenic climate change in the context of the Earth's and mankind's (common) history. This book is full of facts derived from the recent scientific literature, which are woven into the fabric of a gripping narrative. A narrative—given the weight of evidence—that may turn out to have a dark end. As a reader we sit in a space craft watching the planet revolve around the sun. We look 3.5 billion years into the past to follow the evolution of the atmosphere, the origin of life, its rise and decline, and, finally, the most recent appearance of man. Ultimately, Andrew Glikson asks us to ponder the destiny of our own species, which has the intelligence and the ingenuity to put us into space, and also the power to destroy ourselves.

Hans Joachim Schellnhuber
Director of the Potsdam Institute for
Climate Impact Research (PIK)

Preface

This monograph traces milestones in the evolution of the atmosphere, oceans and biosphere from about 3.5 billion years-ago [Ga], through natural cataclysms and all the way to the Anthropocene—a geological era triggered by a species which has uniquely mastered ignition. Of all the factors which have allowed life on Earth one stands out—the presence of liquid water, permitted by the planet's unique orbital position around the sun, its active tectonic and volcanic nature and its evolving atmospheric composition, regulating surface temperatures in the range of ~ -90 to $+58$°C. The atmosphere, mediating the carbon, oxygen and nitrogen cycles, acts as lungs of the biosphere, allowing the existence of an aqueous medium where metabolic microbiological processes occur—from chemo-bacteria around volcanic fumaroles, to nanobes in deep crustal fractures, to near-surface phototrophs. The histories of the atmosphere and of life are thus interdependent. From an initial Venus-like atmosphere dominated by CO_2, CO, SO_2, N_2O, CH_4, H_2 and likely H_2S, the sequestration of CO_2 and the build-up of nitrogen—a stable non-reactive gas—have led to intermittent ice ages from at least as early as ~ 3.0 Ga. This was followed by multi-stage fluctuations in the level of the atmosphere's photosynthetic oxygen produced by phytoplankton and, from ~ 420 Ma, by land plants, ensuing in a flammable carbon-rich biosphere interfaced with an oxygen-rich atmosphere. Superposed on gradual evolutionary trends were abrupt changes triggered by volcanism, asteroid impacts, possible supernovae and episodic release of methane and hydrogen sulphide. Changes in atmospheric chemistry affected variations in alkalinity, acidity [pH] and oxidation/reduction state [Eh] of the hydrosphere and thereby of marine biological processes. Born on a flammable Earth surface, under increasingly unstable climates descending from the warmer Pliocene into the deepest ice ages of the Pleistocene, human survival depended on both—biological adaptations and cultural evolution, mastering fire as a necessity, allowing the species to increase entropy in nature by orders of magnitude. Gathered around camp fires during long nights for hundreds of thousandth of years, captivated by the flickering life-like dance of the flames, humans developed imagination, insights, cravings, fears, premonitions of death and thereby aspiration for immortality, omniscience, omnipotence and the concept of god. Inherent in pantheism was the reverence of the Earth, its rocks and its living creatures, contrasted by the subsequent rise of monotheistic sky-god creeds which regard Earth as but a corridor to heaven. Once the climate stabilized in the early Holocene, since about ~7000 years-ago production of excess food by Neolithic civilization along the Great River Valleys has allowed human imagination and dreams to express themselves through the construction of monuments to immortality. Further to burning large part of the forests, the discovery of combustion and exhumation of carbon from the Earth's ~420 million years-old fossil biospheres set the stage for an anthropogenic oxidation event, affecting an abrupt shift in state of the atmosphere-ocean-cryosphere system. The consequent ongoing extinction equals the past five great mass extinctions of species—constituting a geological event horizon in the history of planet Earth.

Acknowledgments

I wish to thank the following people for their comments and contributions: Wallace Ambrose, Hugh Davies, Colin Groves, Clive Hamilton, Edward Linacre, Tony McMichael, Bruce Radke and Colin Soskolne. Helpful comments were obtained from Leona Ellis, Victor Gostin, Miryam Glikson-Simpson, Barrie Pittock and Reg Morrison. I thank Petra Van Steenbergen and Robert van Gameren for editorial help. I am grateful to Reg Morrison for contributing special figures and photographs, to Jim Gehler for Ediacara photos, Mary White and John Laurie for fossil plant photos and Gerta Keller for reproduction of figures. The following people gave permission to reproduce figures in the book: John Adamek, Anita Andrew, Annemarie Abbondanzo, Robert Berner, Tom Boden, David Bowman, Karl Braganza, Pep Canadell, Randall Carlson, Giuseppe Cortese, Peter deMenocal, Gifty Dzah, Alexey Fedorov, Jim Gehling, Kath Grey, Colin Groves, Jeanette Hammann, James Hansen, Paul Hoffman, John Johnson, Jean Jouzel, Barry Lomax, Petra Löw, Cesca McInerney, Michele McLeod, Yvonne Mondragon, Jennifer Phillips, Miha Razinger, Dana Royer, Elizabeth Sandler, Bill Schopf, Appy Sluijs, Phillipa Uwins, John Valley, Simon Wilde, James Zachos. Brenda Mcavoy kindly helped with proof corrections.

Contents

Part I Early Atmospheres

1 Early Atmosphere-Ocean-Biosphere Systems 3
   References 16

2 Palaeozoic and Mesozoic Atmospheres 21
   References 26

3 Cenozoic Atmospheres and Early Hominins 29
   References 42

Part II The Great Mass Extinctions of Species

4 Mass Extinction of Species 47
   4.1   Acraman Impact and Acritarchs Radiation 49
   4.2   Late Ordovician Mass Extinction 50
   4.3   Late and End-Devonian Mass Extinctions 51
   4.4   Late Permian and Permian–Triassic Mass Extinctions 51
   4.5   End-Triassic Mass Extinction 54
   4.6   Jurassic-Cretaceous Climate Anomalies 54
   4.7   K–T (Cretaceous-Tertiary Boundary) Mass Extinction 54
   4.8   Paleocene-Eocene Extinction 58
   4.9   The End-Eocene Freeze 59
   4.10 Carbon and Oxygen Isotopes and Mass Extinctions 59
   References 65

Part III Homo's Fire Blueprint

5 A Flammable Biosphere 71
   References 74

6	A Fire Species	75
	References	88
7	**Climate and Holocene Civilizations**	91
	References	101

Part IV The Anthropocene Event Horizon

8	*Homo sapiens'* **War Against Nature**	105
	8.1 Neolithic Burning and Early Global Warming	105
	8.2 The Great Carbon Oxidation Event	107
	8.3 The Sixth Mass Extinction of Species	126
	References	129
9	**An Uncharted Climate Territory**	133
	References	147
10	**Homo Prometheus**	149
	References	152

Epilogue: The 'Life Force' 153

Appendices 157

About the Author 163

References 165

Index 167

Part I
Early Atmospheres

we find no vestige of a beginning - no prospect of an end

James Hutton 1785

Chapter 1
Early Atmosphere-Ocean-Biosphere Systems

Abstract The application of isotopic tracers to paleo-climate investigations—including oxygen ($\delta^{18}O$), sulphur ($\delta^{33}S$) and carbon ($\delta^{13}C$), integrated with Sedimentological and proxies studies, allows vital insights into the composition of early atmosphere–ocean-biosphere system, suggesting low atmospheric oxygen, high levels of greenhouse gases (CO_2 + CH_4 and likely H_2S), oceanic anoxia and high acidity, limiting habitats to single-cell methanogenic and photosynthesizing autotrophs. Increases in atmospheric oxygen have been related to proliferation of phytoplankton in the oceans, likely about ~2.4 Ga (billion years-ago) and 0.7–0.6 Ga.

Terrestrial climates are driven through the exposure of the Earth surface to solar insolation cycles (Solanki 2002; Bard and Frank 2006), variations in the gaseous and aerosol composition of the atmosphere, the effects of photosynthesis on CO_2 and O_2 cycles, microbial effects on methane levels, volcanic eruptions, asteroid and comet impacts and other factors. By contrast to the thick CO_2 and SO_2 blankets on Venus, which exert an extreme pressure of 93 bar at the surface, and unlike the thin 0.006 bar atmosphere of Mars, the presence in the Earth's atmosphere of trace concentrations of well-mixed greenhouse gases (GHG) (CO_2, CH_4, N_2O, O_3) modulates surface temperatures, allowing the presence of liquid water and thereby life (Figs. 1.1, 1.2). During the Holocene surface temperatures ranged between −89 and +58 °C, with a mean of about +14 °C.

Forming a thin breathable veneer only slightly more than one thousandths the Earth's diameter, evolving both gradually as well as through major perturbations, the atmosphere acts as lungs of the biosphere, facilitating an exchange of carbon and oxygen with plants and animals (Royer et al. 2004, 2007; Siegenthaler et al. 2005; Berner 2006; Berner et al. 2007; Beerling and Royer 2011). In turn biological activity continuously modifies the atmosphere, for example through production of methane in anoxic environments, release of photosynthetic oxygen from plants and of dimethyl sulfide from marine phytoplankton. Long term chemical changes in the atmosphere-ocean system are affected by changes in plate tectonic-driven geomorphic transformations, including subduction of oceanic and continental plates (Ruddiman 1997, 2003, 2008), weathering, volcanic and methane eruptions, and

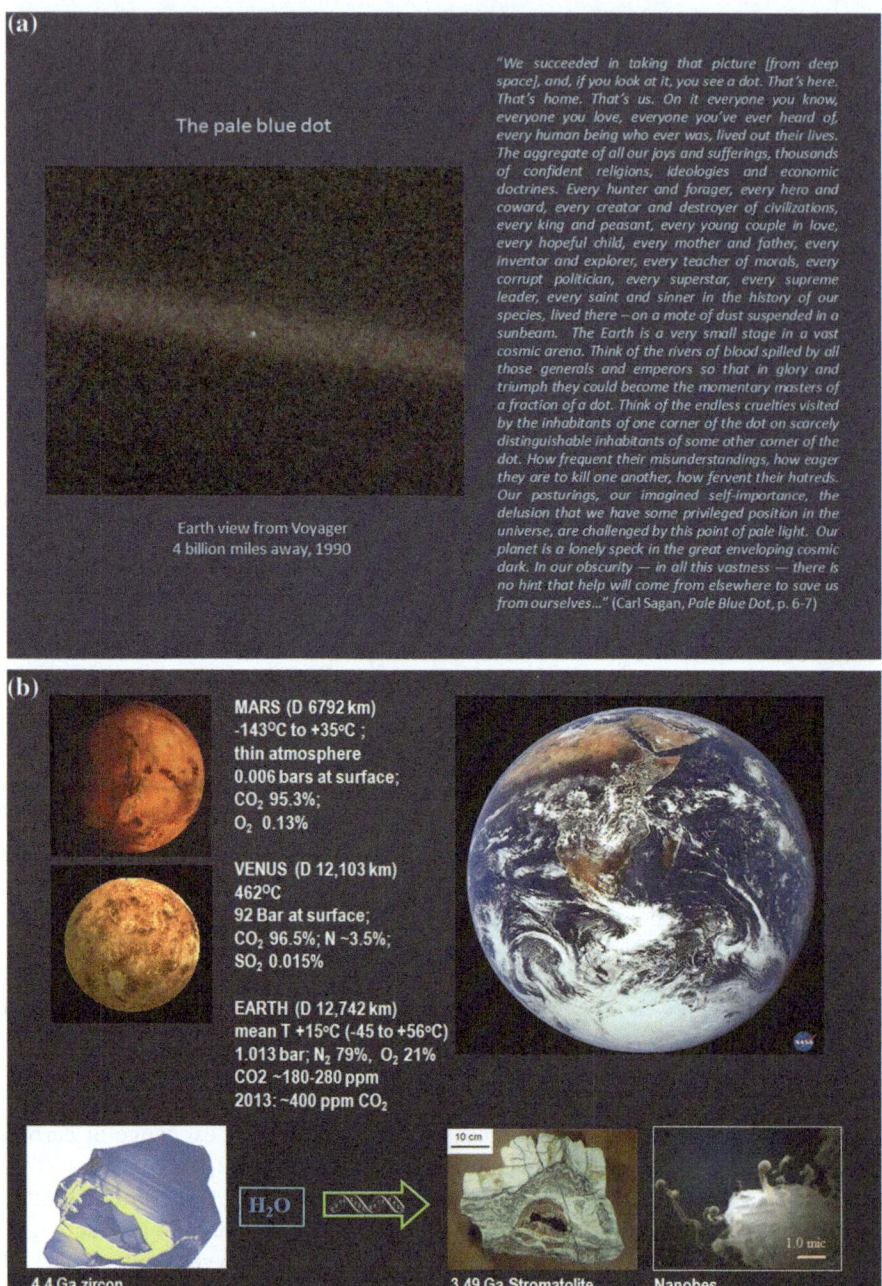

Fig. 1.1 a The pale *blue dot*. Earth viewed by Voyager from a distance of 4 billion miles. The Voyager spacecraft travelled through space at 40,000 miles per hour. NASA http://thornscompose.com/2009/11/28/advent_part1/. **b** Earth-Mars-Venus comparison and the terrestrial evolutionary chain, showing a $\sim$4.4 Ga zircon crystal (Peck et al. 2001, Fig. 8; Elsevier, by permission); water, DNA-RNA chains; Pilbara stromatolite (courtesy J.W. Schopf) and Nanobe organisms (Uwins et al. 1998) (courtesy P. Uwins)

1 Early Atmosphere-Ocean-Biosphere Systems

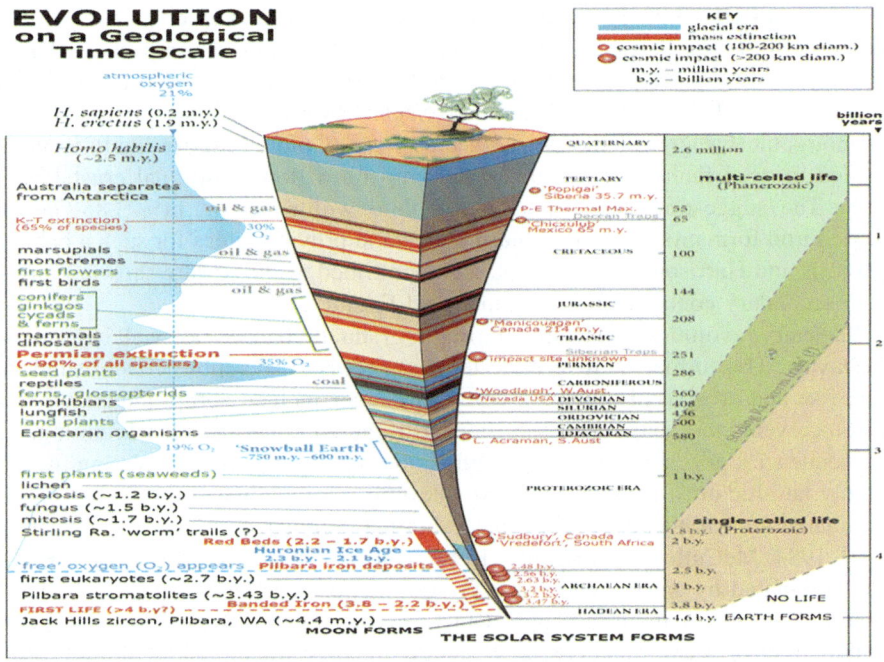

Fig. 1.2 Evolutionary column of life and related events (courtesy Reg Morrison; http://regmorrison.edublogs.org/)

variations in marine and terrestrial photosynthetic activity (Broecker 2000; Zachos et al. 2001; Hansen et al. 2007; Glikson 2008). The range of paleo-climate proxies used in these studies are reviewed in detail by Royer et al. (2001).

Early terrestrial beginnings are interpreted in terms of a cosmic collision ∼4.5 billion years-ago between an embryonic semi-molten Earth and a Mars-scale body—*Theia*—determined from Pb isotopes (Stevenson 1987). The consequent formation of a metallic core, inducing a magnetic field which protects the Earth from cosmic radiation, and a strong gravity field which to a large extent prevents atmospheric gases from escaping into space, resulted in a haven for life at the Earth surface (Gould 1990). Relict ∼4.4 Ga and younger zircons (Fig. 1.1b), representing the *Hadean* era, a term coined by Preston Cloud in 1972, signify vestiges of granitic and felsic volcanic crustal nuclei, implying the presence of a water component in the original magma (Wilde et al. 2001; Mojzsis et al. 2001). Precambrian terrains contain relict ∼4.1–3.8 Ga-old rocks, including volcanic and sedimentary components, exposed in Greenland, Labrador, Slave Province, Minnesota, Siberia, northeast China, southern Africa, India, Western Australia and Antarctica (Van Kranendonk et al. 2007). These formations, formed parallel to the Late Heavy Bombardment (LHB) on the Moon (∼3.95–3.85 Ga) (Ryder 1991), are metamorphosed to an extent complicating recognition of primary impact shock features, which to date precluded an identification of signatures of the LHB on Earth.

During a LHB (3.95–3.85 Ga) exposure of the Earth surface to cosmic and UV radiation from solar flares, incineration affected by large asteroid impacts, and acid rain, are likely to have precluded photosynthesizing bacterial activity (Zahnle and Sleep 1997; Chyba 1993; Chyba and Sagan 1996). However, extremophile chemotrophic bacteria of the *deep hot biosphere* (Gold 1999) are likely to have resided in deep faults and fractures following formation of original crustal vestiges. The suggestion that low $\delta^{13}C$ graphite within apatite in ~3.85 Ga-old banded iron formation in southwestern Greenland provides clues for such a habitat (Mojzsis and Harrison 2000) has not been confirmed as these were shown to arise from secondary contamination (Nutman and Friend 2006).

Planetary evolution transpires through gradual changes as well as major upheavals. The former include plate tectonics, crustal accretion, crustal subduction, rise and erosion of mountain belts. The latter include abrupt magmatic and tectonic events and extra-terrestrial impact-triggered cratering. Geochronological age sequences, geochemistry and isotopic indices point to secular evolution from a mainly basaltic crust (SIMA: Silica-Magnesium-iron-dominated crust) to granite-dominated crustal nuclei (SIAL: Silica-Alumina-dominated crust) (Glikson 1972, 1980, 1984; McCulloch and Bennett 1994). During much of its early history Earth was dominated by an oxygen-poor, CO_2 and methane-rich atmosphere of several thousand to tens of thousands ppm CO_2, resulting in low-pH acid oceans. High temperatures of ocean waters (Knauth and Lowe 2003; Knauth 2005) allowed little sequestration of the CO_2 accumulated in the atmosphere from episodic volcanism, impact cratering, metamorphic release of CO_2, dissociation of methane from sediments and microbial activity and, following the advent of plants on land in the Silurian (~420 Ma), from decomposed vegetation (Berner 2004, 2006; Beerling and Berner 2005; Royer et al. 2004, 2007; Royer 2006; Glikson 2008). The low solubility of CO_2 in the warm water of the early hydrosphere and little weathering-capture of CO_2, due to a low continent/ocean crust ratio and limited exposure of land surface, ensured long term persistence of greenhouse conditions. The development of glaciations, recorded at ~2.9, 2.4–2.2 Ga, 575–543 Ma, the late Ordovician (~446–443 Ma), Carboniferous-Permian (~326–267 Ma), Jurassic (187–163 Ma and post-Eocene, signifies episodic large-scale CO_2 sequestration events (Ruddiman 2003).

During early stages of terrestrial evolution low solar luminosity (*The Faint Young Sun*, 27 % lower luminosity than at present) (Sagan and Mullen 1972), representing a progressive increase in fusion of hydrogen to helium, was compensated by high greenhouse gas (GHG) levels (Fig. 1.3), allowing surface temperature to remain above freezing. Alternative hypotheses were proposed by Longdoz and Francois (1997) in terms of albedo and seasonal variations on the early Earth and by Rosing et al. (2010) in terms of a high ocean/continent surface area ratio in the Archaean, leading to lower albedo and absorption of infrared by open water. Temporal fluctuations in atmospheric GHG levels constituted a major driver of alternating glacial and greenhouse states (Kasting and Ono 2006). Knauth and Lowe (2003) and Knauth (2005) measured low $\delta^{18}O$ values in ~3.5–3.2 Ga cherts of the Onverwacht Group, Barberton Greenstone Belt (BGB), Kaapvaal

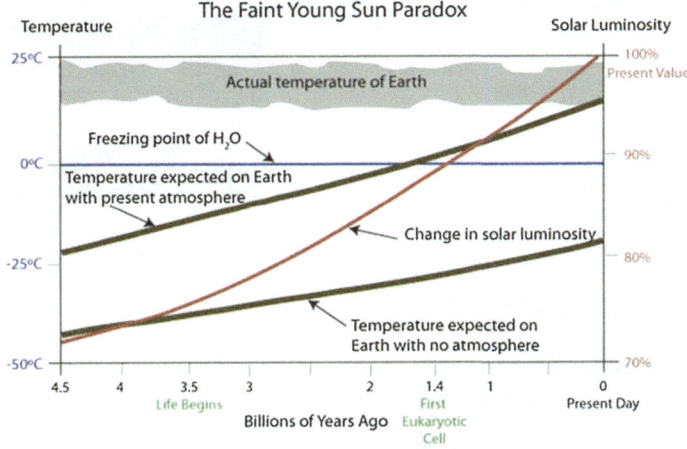

Fig. 1.3 The faint young sun paradox according to Sagan and Mullen (1972), suggesting compensation of the lower solar luminosity by high atmospheric greenhouse gas levels at early stages of terrestrial evolution. (From the *Habitable Planet*: A system approach to environmental science, produced by the Harvard-Smithsonian Center for Astrophysics, Science Media Group and used with permission by the Annenberg Learner (Courtesy Michele McLeod) www.learner.org; http://www.learner.org/courses/envsci/unit/text.php?unit=1&secNum=4)

Craton, suggesting extremely high ocean temperatures in the range of 55–85 °C (Fig. 1.4a). The maximum $\delta^{18}O$ value in Barberton chert (+22 ‰) is lower than the minimum values (+23 ‰) in Phanerozoic sedimentary cherts, precluding late diagenesis as the explanation of the overall low $\delta^{18}O$ values. Regional metamorphic, hydrothermal, or long-term resetting of original $\delta^{18}O$ values is also precluded by preservation of $\delta^{18}O$ across different metamorphic grades. According to Knauth (2005) high-temperature conditions extended beyond submarine fumaroles and the Archaean oceans were characterized by high salinities × 1.5–2.0 times the modern level. In this interpretation ensuing evaporite deposits were removed by subduction, allowing lower salinities. However, well preserved Archaean sedimentary sequences contain little evidence of evaporite deposits. The low-oxygen levels of the Archaean atmosphere and hydrosphere limited marine life to extremophile cyanobacteria. Microbial methanogenesis involves reactions of CO_2 with H_2 and in organic molecules produced from fermentation of photosynthetically produced organic matter. Photolysis of methane may have created a thin atmospheric organic haze.

An overall increase with time in $\delta^{18}O$ shown by terrestrial sediments (Valley 2008) (Fig. 1.4b) reflects a long term cooling of the hydrosphere, consistent with an overall but intermittent temporal decline in atmospheric CO_2 shown by plant leaf pore (stomata) studies (Berner 2004, 2006; Beerling and Berner 2005; Royer et al. 2001, 2004, 2007) (Fig. 2.1). This long-term decline may have been associated with increased rates of weathering-sequestration of CO_2 related to erosion

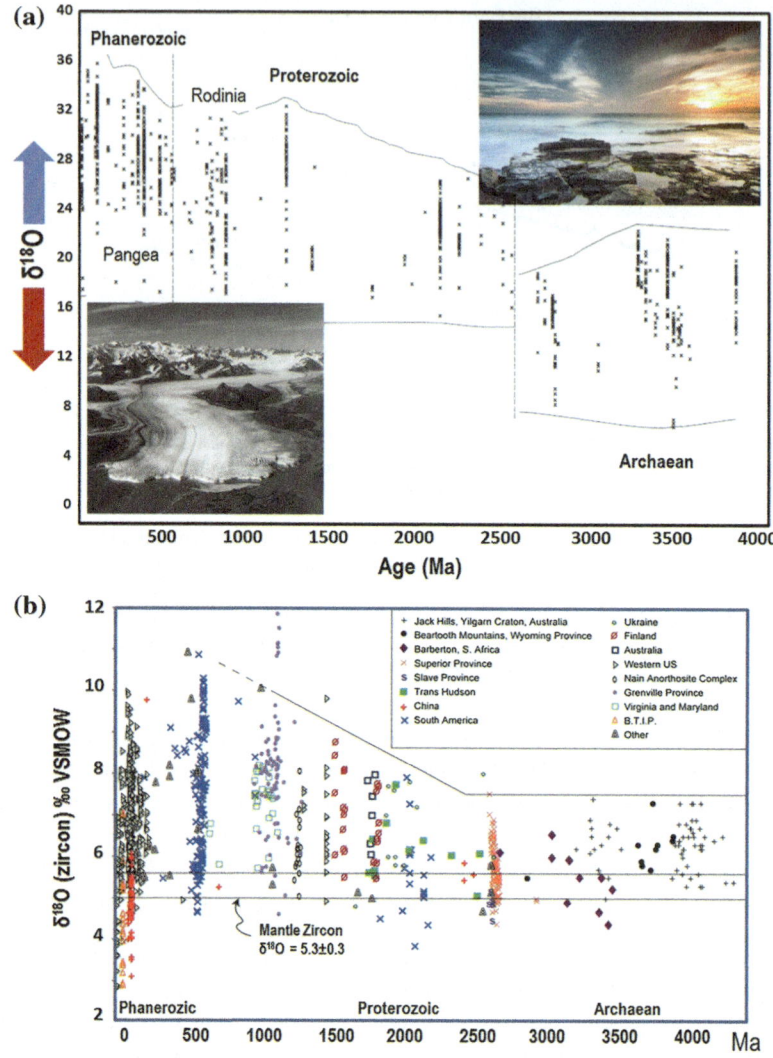

Fig. 1.4 a A compilation of oxygen isotope data for cherts. The overall increase in $\delta^{18}O$ with time is interpreted as global cooling over the past 3,500 Ma. The variation in $\delta^{18}O$ for cherts at any given time is caused by the presence of low $\delta^{18}O$ meteoric waters during burial at elevated temperatures (Knauth 2005, Fig. 1; Elsevier, by permission). **Inset 1**: Columbia Glacier (NASA) (http://www.google.com.au/search?q=nasa+glacier&hl=en&tbm=isch&tbo=u&source=univ&sa=X&ei=hkmEUbCDBonGkgXDg4C4CQ&sqi=2&ved=0CEYQsAQ&biw=1360&bih=878); **Inset 2**. anieto2k's photo stream. http://www.flickr.com/photos/anieto2k/8636213185/sizes/l/in/photostream/ Creative Commons: http://creativecommons.org/licenses/by-sa/2.0/. **b** $\delta^{18}O$ ratio of igneous zircons from 4.4 Ga to recent, displaying an increase in abundance of low-temperature effects with time from approximately ~2.3 Ga (Courtesy J.W. Valley)

1 Early Atmosphere-Ocean-Biosphere Systems

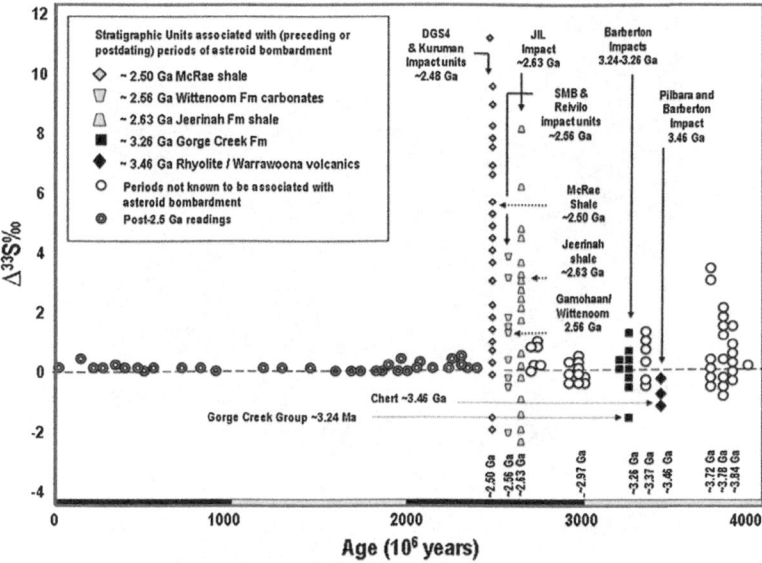

Fig. 1.5 Plots of mass independent fractionation values for Sulphur isotopes ($\Delta^{33}S$ − MIF-S) vs Age. The high $\Delta^{33}S$ values for pre-2.45 Ga sulphur up to about $\Delta^{33}S = 11$ is interpreted in terms UV-induced isotopic fractionation, allowed by a lack of an ozone layer. Periods of major asteroid impacts during which ozone may have been destroyed are indicated (from Glikson 2010, Fig. 1; Elsevier, by permission)

of rising orogenic belts, including the Caledonian, Hercynian, Alpine, Himalayan and Andean mountain chains (Ruddiman 1997, 2003). Such a trend is consistent with suggested increase in the role plate tectonics through time (Glikson 1980), which led to an increase in sequestration of CO_2 by weathering of uplifted orogenic belts and subduction of CO_2-rich carbonate and carbonaceous shale.

Central to studies of early atmospheres is the level of oxygen and its relation to photosynthesis. Sulphur isotopic analyses record mass-independent fractionation of sulphur isotopes ($\delta^{33}S$) (MIF-S) in sediments older than ∼2.45 Ga, widely interpreted in terms of UV-triggered reactions under oxygen-poor ozone-depleted atmosphere and stratosphere (Farquhar et al. 2000, 2007) (Fig. 1.5). From about ∼2.45–2.32 Ga—a period dominated by deposition of banded iron formations (BIF) (Fig. 1.6), $\delta^{33}S$ values signify development of an ozone layer shielding the surface from UV radiation (Farquhar et al. 2000; Kump 2009) (Fig. 1.5). According to Kopp et al. (2005) photosynthetic oxygen release from cyanobacteria and aerobic eukaryotes affected oxidation of atmospheric methane, triggering planetary-scale glaciation at least as early as 2.78 Ga and perhaps as long ago as 3.7 Ga. According to Ohmoto et al. (2006) MIF-S values and oxygen levels fluctuated through time and place, possibly representing local volcanic activity. The role of asteroid impacts in affecting the ozone layer, UV radiation and thus regional MIF-S values has been discussed by Glikson (2010).

Fig. 1.6 Banded iron formations (BIF). **a** Outcrops of ~2.48 Ga BIF of Hamersley Group, Pilbara Craton, Western Australia, at Dales Gorge near Fortescue Fall; **b** Magnetite-quartz banded iron formation, Isua Supracrustal belt, Southwest Greenland. Reflected light in thin section

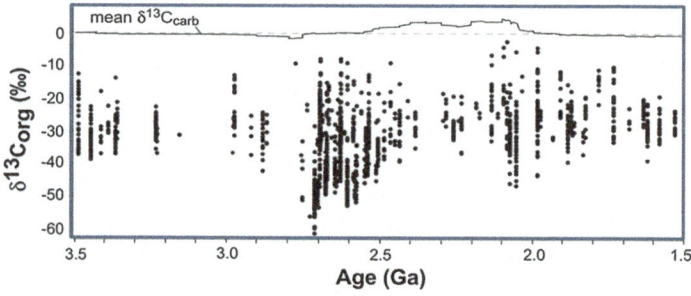

Fig. 1.7 A compilation of published kerogen and total organic carbon δ^{13}C values (δorg) for all sedimentary rock types. Note high accumulations of organic carbon in the late Archaean at ~2.7–2.6 Ga. (Eigenbrode and Freeman 2006, Fig. 1; Proceedings of the National Academy of Science by permission)

Some of the earliest manifestations of biological activity may be represented by banded iron formations (BIF) from the ~3.8 Ga Isua supracrustal belt (southwest Greenland) (Fig. 1.6). Banded iron formations are commonly intercalated with

1 Early Atmosphere-Ocean-Biosphere Systems

Fig. 1.8 Stromatolites—early contributors to atmospheric oxygen. **a** ~3.49 Ga Stromatolites colony, Dresser Formation, North Pole dome, central Pilbara, Western Australia (Courtesy JW Schopf); **b** ~ 3.49 Ga stromatolite (location as for **a**); **c** 3.43 Ga stromatolites, Pilbara Craton, Western Australia (photograph—courtesy Reg Morrison); **d** ~3.43 Ga stromatolites, Pilbara Craton, Western Australia; **e** ~2.76 Ga stromatolite, Meentheena Formation, Nullagine River, Western Australia; **f** ~2.63 Ga mega-stromatolite, Carawine Pool, Oakover River, Eastern Pilbara, Western Australia (*arrow points* to the *upper left* rim of the stromatolite dome). (Photos D, E, F by the author); **g** Living stromatolites, Shark Bay, Western Australia (Photograph—Courtesy Reg Morrison)

volcanic tuff and carbonaceous shale whose low $\delta^{13}C$ indices are indicative of biological activity. The carbon isotopic compositions of Archaean black shale, chert and BIF provide vital clues to the proliferation of autotrophs in the shallow and deep marine environment. Peak biogenic productive periods about 2.7–2.6 Ga are represented by low $\delta^{13}C$ of chert and black shale intercalated with banded iron formations in the Superior Province, Canada (Goodwin et al. 1976), and in the Hamersley Basin, Western Australia (Eigenbrode 2006) (Fig. 1.7). This peak, which coincides with intense volcanic activity in greenstone belts world-wide, suggests enhanced biological activity related to volcanic emanations and enriched nutrient supply. Biological processes include oxygen capture by iron-oxidizing microbes, microbial methanogenesis producing atmospheric CH_4, microbial sulphur metabolism producing H_2S, ammonia-releasing microbes, oxygen-releasing photosynthesizing colonial prokaryotes (stromatolites) (Fig. 1.8) and algae, culminating in production of O_2-rich atmosphere and the O_3 ozone layer. Earliest manifestations of biological forcing may be represented by banded iron formations, widely held to represent ferrous to ferric iron oxidation by microbial reactions (Cloud 1968, 1973; Morris 1993; Konhauser et al. 2002; Glikson 2006).

The origin of BIFs has been interpreted in terms of transportation of ferrous iron in ocean water under oxygen-poor atmospheric and hydrospheric conditions of the Archaean (Cloud 1968; Morris 1993). Oxidation of ferrous to ferric iron could occur through chemotrophic or phototrophic bacterial processes (Konhauser et al. 2002) and/or by UV-triggered photo-chemical reactions. The near-disappearance of banded iron formations (BIF) about ~2.4 Ga (Fig. 1.6), with transient reappearance about 1.85 Ga and in the Vendian (650–543 Ma), likely reflect increase in oxidation, where ferrous iron became unstable in water and the deposition of BIF was replaced by that of detrital hematite and goethite. Archaean impact ejecta units in the Pilbara and Kaapvaal Cratons are commonly overlain by ferruginous shale and BIF (Glikson 2006; Glikson and Vickers 2007), hinting at potential relations between Archaean impact clusters, impact-injected sulphate, consequent ozone depletion, enhanced UV radiation and formation of BIFs (Glikson 2010), possibly by photolysis through the reaction by photolysis.

Some of the oldest possible micro-fossils occur in black chert of the Dresser Formation (Duck et al. 2008; Golding and Glikson 2011) and in brecciated chert of the ~3,465 Ma Apex Basalt, Warrawoona Group, Pilbara Craton (Schopf et al.

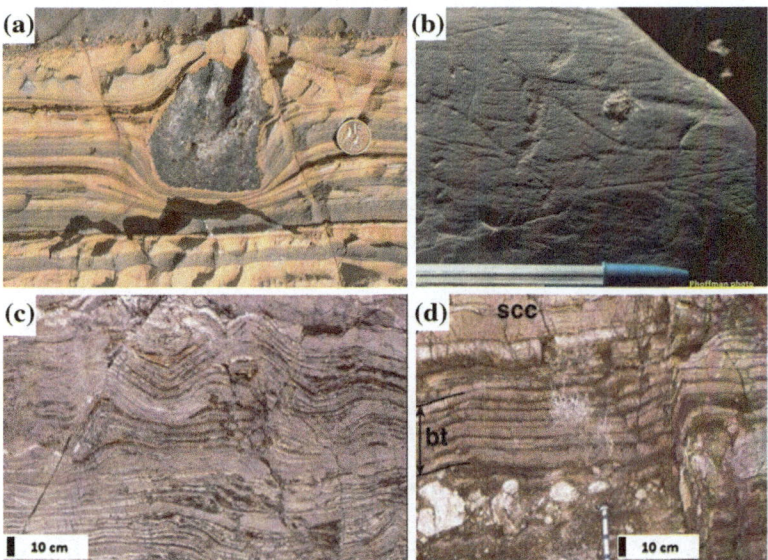

Fig. 1.9 Evidence of the Cryogenian 'snow-ball' Earth (750–635 Ma); **a** example of ice-rafted dropstone in pro-glacial marine strata, Ghaub Formation, Otavi Group, Namibia (Courtesy Paul Hoffman); **b** Scratched pebble from Jbeliat tillite, Mauritania (Courtesy Paul Hoffman); **c** Cap carbonates, Namibia—mechanically-laminated peloidal dolostone (Hoffman et al. 2007, Fig. 8, Elsevier, with permission); **d** Cap carbonates, Namibia—peloidal dolostone (Hoffman et al. 2007, Fig. 8; Elsevier, with permission)

2007). The paleo-environment, carbonaceous composition, mode of preservation, and morphology of these microbe-like filaments, backed by new evidence of their cellular structure provided by two- and three-dimensional Raman imagery, support a biogenic interpretation. Evidence for hydrothermal and methanogenic microbial activity (Schopf and Packer 1987; Schopf et al. 2007; Hoffman et al. 1999; Duck et al. 2008; Golding and Glikson 2011) and intermittent appearance of shallow water stromatolites in $\sim$3.49 Ga (Fig. 1.8a; Dunlop and Buick 1981) and $\sim$3.42 Ga sediments (Fig. 1.8c; Alwood et al. 2007) testify to a diverse microbial habitat. This included heliotropic and by implication photosynthesizing stromatolites affecting release of oxygen. Problems in identifying early Archaean stromatolites were expressed by Lowe (1994) and by Brazier et al. (2002). Whereas the early stromatolites may represent Prokaryotes, Eukaryotes possibly appeared about $\sim$2.1–1.6 Ga (Knoll et al. 2006), or earlier (Sugitania et al. 2009) and exist at present at Shark Bay, Western Australia (Fig. 1.8g).

The occurrence of glacial stages during the Precambrian (pre-0.54 Ga) despite high atmospheric greenhouse gas levels is accounted for by *Faint Early Sun* conditions (Fig. 1.3), examples include:

(a) An upper Archaean glaciation represented by the 5,000 m-thick Mozaan Group of the 2,837 ± 5 Ma Pongola Supergroup (Strik et al. 2007), which includes a sequence of diamictite containing striated and faceted clasts and ice-rafted debris (Young et al. 1998), representing the oldest glaciation recorded to date.
(b) An early Proterozoic Huronian glaciation (~2.4–2.2 Ga) recorded by outcrops in the North American Great Lakes district, Pilbara Western Australia and the Transvaal (Kopp et al. 2005).

Fig. 1.10 Ediacara fossils, Flinders Ranges, South Australia; *1* Dickinsonia *costata*; *2* Parvancorina *minchami*; *3* Sprigina *holotype*; *4* Tribrachidium *heraldicum*; *5* Eoandromeda *octobrachiata* (Courtesy Jim Gehling, South Australia Museum)

1 Early Atmosphere-Ocean-Biosphere Systems

Fig. 1.11 Cambrian arthropods. *1* Leanchoilia *superlata*; *2* Bergeroniellus *asiaticus*, *family Redlichiidae*; *3* Ptychagnostus *atavus*; *4* Marella *splendens*; *5* Marella *splendens* (Courtesy of John P Adamek, Fossilmuseum; http://www.fossilmuseum.net/)

(c) The late Proterozoic Cryogenian glaciations (*snowball Earth*), including the Sturtian (∼750–700 Ma) and Marinoan (∼650–635 Ma) (Fig. 1.9), which affected an extensive ocean-wide ice cover, preventing sequestration of volcanic-emitted CO_2 and leading to an atmospheric build-up of the gas to extreme levels, some × 350 times the modern concentration (Hoffman et al. 1998). The ice prevented CO_2 sequestration, resulting in a powerful greenhouse effect which culminated in glacial collapse represented by an extensive *cap carbonate* unit which overlies fragmented glacial deposits (Hoffman et al. 1998; Hoffman and Schrag 2000, 2002; Halverson et al. 2005).

The glaciation, cooling of the oceans and enrichment of oxygen in cold water led to enhanced photosynthesis by phytoplankton and thereby further enrichment of oxygen. Likewise Kasting and Ono (2006) invoke biological activity as driver of the ∼2.4 Ga glaciation, including a photosynthetic rise in O_2 and concomitant decrease in CH_4.

Glacial deposits of the Cryogenian Snowball Earth (750–635 Ma) observed in Namibia, in South Australia, Oman and Svalbard, correspond to the period of fragmentation of the long-lived Rodinia supercontinent (∼1.1–0.75 Ga) (Hoffman et al. 1998) (Fig. 1.9). The onset of glaciation as marked by a large negative $\delta^{13}C$ anomaly and glacial termination as marked by carbonates—the so-called *cap carbonate* (Halverson et al. 2005). Paleomagnetic evidence suggests that the ice sheets reached sea level close to the equator during at least two glacial episodes.

Some glacial units include sedimentary iron formations, underpinning potential relations between BIF and glaciations. According to Kirschvink (1992) the runaway albedo feedback exerted by the ice sheets resulted in a near-global ocean ice cover whereas continental ice covers were thin due to retardation of the hydrological cycle. In this model the appearance of banded iron formations may represent sub-glacial anoxia and thereby enrichment of sea water in ferrous iron. Hoffman et al. (1998) report Negative carbon isotope anomalies in carbonate rocks bracketing Neoproterozoic glacial deposits in Namibia, indicating collapse of oceanic biological activity over millions of years, accounted for by a global glaciation. Glaciation ended abruptly in connection with volcanic outgassing, raising atmospheric CO_2 to some $\times$ 350 times the modern level.

The late Proterozoic thus represents a transition from oxygen-poor composition of early atmospheres which were dominated by reduced carbon species in the air and the oceans, producing methane through inorganic, organic and microbial processes by chemo-bacteria. Following the Cryogenian ice age $\sim$750–635 Ma (Hoffman et al. 1998; Hoffman and Schrag 2000) the rise in oxygen during $\sim$635–542 Ma and particularly following $\sim$580 Ma allowed oxygen-binding proteins and emergence of the multicellular Ediacara fauna (Fig. 1.10) in an oxygenated *Canfield Ocean*. According to Canfield et al. (2007) oxygen levels constituted the critical factor allowing multicellular animals to emerge in the late-Neoproterozoic, as evidenced from the oxidation state of iron before and after the Cryogenian glaciation. A prolonged stable oxygenated environment may have permitted the emergence of bilateral motile animals some 25 million years following glacial termination, later followed by the onset of the *Cambrian explosion* of life (Gould 1990) from $\sim$542 Ma and development of a rich variety of organisms (Fig. 1.11).

References

Allwood AC, Walter MR, Burch IW, Kamber BS (2007) 343 billion-year-old stromatolite reef from the Pilbara Craton of Western Australia: ecosystem-scale insights to early life on. Earth Precam Res 158:198–227

Bard E, Frank M (2006) Climate change and solar variability: what's new under the sun? Earth Planet Sci Lett 248:1–14

Beerling DJ, Berner RA (2005) Feedbacks and the coevolution of plants and atmospheric CO_2. Proc Nat Acad Sci 102:1302–1305

Beerling DJ, Royer D (2011) Convergent Cenozoic CO_2 history. Nat Geosci 4:418–420

Berner RA (2004) The phanerozoic carbon cycle: CO_2 and O_2. Oxford University Press, New York

Berner RA (2006) GEOCARBSULF: a combined model for Phanerozoic atmospheric O_2 and CO_2. Geochim et Cosmochim Acta 70:5653–5664

Berner RA, Vanderbrook JM, Ward PD (2007) Oxygen and evolution. Science 316:557–558

Brazier MD, Green OR, Jephcoat AP, Kleppe AK, Van Kranendonk MJ, Lindsay JF, Steele A, Grassineau NV (2002) Questioning the evidence for Earth's oldest fossils. Nature 416:76–81

References

Broecker WS (2000) Abrupt climate change: causal constraints provided by the paleoclimate record. Earth Sci Rev 51:137–154

Canfield D, Poulton SW, Narbonne GM (2007) Late-neoproterozoic deep-ocean oxygenation and the rise of animal life. Science 315:92–95

Chyba CF (1993) The violent environment of the origin of life: progress and uncertainties. Geochim et Cosmochim Acta 57:3351–3358

Chyba CF, Sagan C (1996) Comets as the source of prebiotic organic molecules for the early Earth. In: Thomas PJ, Chyba CF, McKay CP (eds) Comets and the origin and evolution of life. Springer, New York, pp 147–174

Cloud P (1968) Atmospheric and hydrospheric evolution of the primitive. Earth Sci 160:729–738

Cloud P (1973) Paleoecological significance of the banded iron formation. Econ Geol 68:1135–1143

Duck LJ, Glikson M, Golding SD, Webb R, Riches J, Baiano J, Sly L (2008) Geochemistry and nature of organic matter in 35 Ga rocks from Western Australia. Geochim Cosmochim Acta 70:1457–1470

Dunlop JSR, Buick R (1981) Archaean epiclastic sediments derived from mafic volcanics, North Pole, Pilbara Block, Western Australia. Geol Soc Aust 7:225–233

Eigenbrode JL, Freeman KH (2006) Late Archaean rise of aerobic microbial ecosystems. Proc Nat Acad Sci 103:15759–15764

Farquhar J, Bao H, Thiemens M (2000) Atmospheric influence of Earth's earliest sulfur cycle. Science 289:756

Farquhar J, Peters M, Johnston DT, Strauss H, Masterson A, Wiechert U, Kaufman AJ (2007) Isotopic evidence for Mesoarchaean anoxia and changing atmospheric sulphur chemistry. Nature 449:706–709

Glikson AY (1972) Early precambrian evidence of a primitive ocean crust and island nuclei of sodic granite. Geol Soc Am Bull 83:3323–3344

Glikson AY (1980) Uniformitarian assumptions, plate tectonics and the Precambrian Earth. In: Kroner A (ed) Precambrian plate tectonics. Elsevier, Amsterdam, pp 91–104

Glikson AY (1984) Significance of early Archaean mafic–ultramafic xenolith patterns. In: Kroner A, Goodwin AM, Hanson GN (eds) Archaean geochemistry. Springer, Berlin, pp 263–280

Glikson AY (2006) Asteroid impact ejecta units overlain by iron-rich sediments in 3.5–2.4 Ga terrains, Pilbara and Kaapvaal cratons: Accidental or cause–effect relationships? Earth Planet Sci Lett 246:149–160

Glikson AY (2008) Milestones in the evolution of the atmosphere with reference to climate change. Aust J of Earth Sci 55:125–139

Glikson AY (2010) Archaean asteroid impacts, banded iron formations and MIF-S anomalies: a discussion. Icarus 207:39–44

Glikson AY, Vickers J (2007) Asteroid mega-impacts and Precambrian banded iron formations: 2.63 Ga and 2.56 Ga impact ejecta/fallout at the base of BIF/argillite units, Hamersley Basin, Pilbara Craton. Western Australia. Earth Planet Sci Lett 254:214–226

Golding S, Glikson MV (2011) Earliest life on earth: habitats, environments and methods of detection. Springer, Dordrecht

Gold T (1999) The deep hot biosphere. Springer, New York, p 235

Goodwin AM, Monster J, Thode HG (1976) Carbon and sulfur isotope abundances in Archean iron-formations and early Precambrian life. Econ Geol 71:870–891

Gould SJ (1990) Wonderful life: the burgess shale and the nature of history. W W Norton and Company Inc, New York, p 347

Halverson GP, Hoffman PF, Schrag DP, Maloof AC, Adam C, Hugh A, Rice N (2005) Toward a Neoproterozoic composite carbon-isotope record. GSA Bull 117:1181–1207

Hansen J, Sato M, Kharecha P, Lea DW, Siddall M (2007) Climate change and trace gases. Phil Trans Roy Soc 365A:1925–1954

Hoffman PF, Schrag DP (2000) Snowball Earth. Sci Am 282:68–75

Hoffman PF, Schrag DP (2002) The snowball Earth hypothesis: testing the limits of global change. Terra Nova 14:129–155

Hoffman PF, Kaufman AJ, Halverson GP, Schrag DP (1998) A neoproterozoic snowball Earth. Science 281:1342–1346
Hoffman PF, Halverson GP, Domack JM, Husson JA, Higgins D, Schrag DP (2007) Are basal Ediacaran (635 Ma) post-glacial "cap dolostones" diachronous? Earth Planet Sci Lett 258:114–131
Hofmann HJ, Grey K, Hickman AH, Thorpe RI (1999) Origin of 3.45 Ga Coniform Stromatolites in the Warrawoona Group, Western Australia. Bull Geol Soc Am 111:1256–1262
Kasting JF, Ono S (2006) Palaeoclimates: the first two billion years. Philos Trans R Soc Biol Sci 361:917–929
Kirschvink JL (1992) In: Schopf JW, Klein C (eds.) The proterozoic biosphere. Cambridge Univ Press, New York, p 51–52
Knauth LP (2005) Temperature and salinity history of the Precambrian ocean: implications for the course of microbial evolution. Palaeogeo Palaeoclimat Palaeoecol 219:53–69
Knauth LP, Lowe DR (2003) High Archaean climatic temperature inferred from oxygen isotope geochemistry of cherts in the 3.5 Ga Swaziland Supergroup, South Africa. GSA Bulletin 115(5):566–580
Knoll AH, Javaux EJ, Hewitt D, Cohen P (2006) Eukaryotic organisms in Proterozoic oceans. Phil Trans R Soc London Part B 361:1023–1038
Konhausser K, Hamada T, Raiswell R, Morris R, Ferris F, Southam G, Canfield D (2002) Could bacteria have formed the Precambrian banded iron-formations? Geology 30:1079–1082
Kopp RE, Kirschvink JL, Hilburn IA, Nash CZ (2005) The Paleoproterozoic snowball Earth: a climate disaster triggered by the evolution of oxygenic photosynthesis. Proc Nat Acad Sci 102:11131–11136
Kump LR (2009) The rise of atmospheric oxygen. Nature 451:277–278
Longdoz B, Francois LM (1997) The faint young sun climatic paradox: influence of the continental configuration and of the seasonal cycle on the climatic stability. Global Planet Change 14:97–112
Lowe DR (1994) Abiological origin of described stromatolites older than 3.2 Ga. Geology 22:387–390
McCulloch MT, Bennett VC (1994) Progressive growth of the Earth's continental crust and depleted mantle: geochemical constraints. Geochim Cosmochim Acta 58:4717–4738
Mojzsis SJ, Harrison TM (2000) Vestiges of a beginnings: clues to the emergent biosphere recorded in the oldest known rocks. GSA Today 10:1–6
Mojzsis SJ, Harrison TM, Pidgeon RT (2001) Oxygen-isotope evidence from ancient zircons for liquid water at the Earth's surface 4,300 Myr ago. Nature 409:178–181
Morris RC (1993) Genetic modeling for banded iron-formation of the Hamersley Group, Pilbara Craton, Western Australia. Precamb Res 60:243–286
Nutman AP, Friend CRL (2006) Re-evaluation of oldest life evidence: Infrared absorbance spectroscopy and petrography of apatites in ancient metasediments, Akilia, W. Greenland. Precamb Res 147:100–106
Ohmoto H, Watanabe Y, Ikemi H, Poulson SR, Taylor BE (2006) Sulphur isotope evidence for an oxic Archaean atmosphere. Nature 442:908–911
Peck WH, Valley JW, Wilde SA, Graham CM (2001) Oxygen isotope ratios and rare earth elements in 3.3 to 4.4 Ga zircons: Ion microprobe evidence for high $\delta^{18}O$ continental crust and oceans in the Early Archaean. Geochim et Cosmochim Acta 65:4215–4229
Rosing MT, Bird DK, Sleep NH, Bjerrum CJ (2010) No climate paradox under the faint early Sun. Nature 464:744–749
Royer DL (2006) CO_2-forced climate thresholds during the Phanerozoic. Geochim Cosmochim Acta 70:5665–5675
Royer DL, Berner RA, Beerling DJ (2001) Phanerozoic atmospheric CO change: evaluating geochemical and paleobiological approaches. Earth Sci Rev 54:349–392
Royer DL, Berner RA, Montañez I, Neil P, Tabor J, Beerling DJ (2004) CO_2 as a primary driver of Phanerozoic climate. GSA Today 14:3

References

Royer DL, Berner RA, Park J (2007) Climate sensitivity constrained by CO_2 concentrations over the past 420 million years. Nature 446:530–532

Ruddiman WF (1997) Tectonic uplift and climate change. Plenum Press, New York, p 535

Ruddiman WF (2003) Orbital insolation, ice volume, and greenhouse gases. Quatern Sci Rev 22:1597–1629

Ruddiman WF (2008) Earth's climate, past and future (2nd edn). WH Freeman ISBN 978-0-7167-8490-6

Ryder G (1991) Accretion and bombardment in the Earth–Moon system: the Lunar record. Lunar Planet Sci Instit Contrib 746:42–43

Sagan C, Mullen G (1972) Earth and mars: evolution of atmospheres and surface temperatures. Science 177:52–56

Schopf JW, Packer BM (1987) Early Archean (3.3-billion to 3.5-billion-year-old) microfossils from Warrawoona Group, Australia. Science 237:70–73

Schopf JW, Kudryavtsev AB, Czaja AD, Tripathi AB (2007) Evidence of Archean life: stromatolites and microfossils. Precamb Res 158:141–155

Siegenthaler U et al (2005) Stable carbon cycle–climate relationship during the late pleistocene. Science 310:1313–1317

Solanki SK (2002) Solar variability and climate change: is there a link? Sol Phys 43:59–513

Stevenson DJ (1987) Origin of the Moon-the collision hypothesis. Ann Rev Earth Planet Sci 15:271–315

Strik G, de Wit MJ, Langereis CG (2007) Palaeomagnetism of the Neoarchaean Pongola and Ventersdorp supergroups and an appraisal of the 30–19 Ga apparent polar wander path of the Kaapvaal Craton, Southern Africa. Precamb Res 153:96–115

Sugitania K, Grey K, Nagaokac T, Mimurad K, Walter M (2009) Taxonomy and biogenicity of Archaean spheroidal microfossils (ca 3.0 Ga) from the Mount Goldsworthy–Mount Grant area in the northeastern Pilbara Craton, Western Australia. Precamb Res 173:50–59

Uwins PJR et al (1998) Novel nano-organisms from Australian sandstones. Am Mineral 83:1541–1550

Valley JW (2008) The origin of habitats. Geology 36:911–912

Van Kranendonk MJ (2007) Tectonics of the early Earth. In: Van Kranendonk MJ, Smithies RH, Bennett VC (eds) Earth's oldest rocks, developments in precambrian geology, vol 15. Elsevier, Amsterdam, p 1105–1116

Wilde SA, Valley JW, Peck WH, Graham CM (2001) Evidence from detrital zircons for the existence of continental crust and oceans on the Earth 4.4 Gyr ago. Nature 409:175–178

Young GM, von Brunn V, Gold WEL, Minter DJC (1998) Earth's oldest reported glaciation: physical and chemical evidence from the Archean Mozoan Group (∼2.9 Ga). S Africa J Geol 106:523–538

Zachos J, Pagani M, Sloan L, Thomas E, Billups K (2001) Trends, rhythms, and aberrations in global climate 65 Ma to present. Science 292:686–693

Zahnle K, Sleep NH (1997) Impacts and the early evolution of life. In: Thomas PJ, Chyba CF, McKay CP (eds) Comets and the origin and evolution of life. Springer, New York, p 175–208

Chapter 2
Palaeozoic and Mesozoic Atmospheres

Abstract Detailed investigations of the carbon, oxygen and sulphur cycles using a range of proxies, including leaf pore stomata, $\delta^{13}C$, $\delta^{34}S$, and $^{87/86}Sr$ isotopes, as well as geochemical mass balance modeling, provide detailed evidence of major trends as well as distinct events in the atmosphere-ocean-land system during the Paleozoic and Mesozoic eras (542–65 Ma), including greenhouse Earth periods ($CO_2 \sim 2{,}000$–$5{,}000$ ppm) and glacial phases ($CO_2 < 500$ ppm), with implications for biological evolution.

Phanerozoic geochemical models of the carbon, oxygen and sulphur cycles by Berner et al. (2007) underpin the production of O_2 through weathering and burial of organic carbon and pyrite, while biological productivity is indicated by $^{13}C/^{12}C$ and $^{34}S/^{32}S$ indices. These studies reveal well pronounced trends consistent with geological and paleontological observations (Beerling et al. 2001; Berner 2004, 2005, 2006, 2009; Berner et al. 2007; Beerling and Royer 2011) (Figs. 2.1, 2.2, 3.1, 3.2). Investigations of the carbon isotope composition of Paleozoic organic matter, including plant fossils, coal, bulk terrestrial organic matter and carbonates ($\delta^{13}C_{\text{Total Organic Matter}}$, $\delta^{13}C_{\text{carbonate}}$, $\delta^{13}C_{CO2}$) confirm the dependence of both organic and inorganic carbon composition on the atmospheric O_2/CO_2 ratios (Strauss and Peters-Kottig 2003). The data underpin the intertwined nature of the carbon cycle, oxygen cycle and biological productivity, where the fractionation of atmospheric CO_2 through photosynthesis represents the relations between global photosynthesis (CO_2 capture, O_2 production) and global respiration (O_2 capture; CO_2 release and burial).

Throughout the Phanerozoic periods of low atmospheric CO_2 coincided with periods of high atmospheric O_2 and high biological productivity, examples being the Carboniferous-Permian interval (326–267 Ma) and upper Cenozoic post-32 Ma glaciations (Figs. 2.1, 2.2). During glacial periods a large fraction of CO_2 is sequestered by cold ocean waters and fractionation of atmospheric CO_2 occurs via photosynthesis, where the carbon-rich residues of land plants and phytoplankton are buried and removed from the atmosphere-biosphere system, whereas oxygen is released back to the atmosphere. During high-CO_2 periods the warmer ocean waters sequester less CO_2 and dissolve less oxygen. CO_2 saturation of warm water

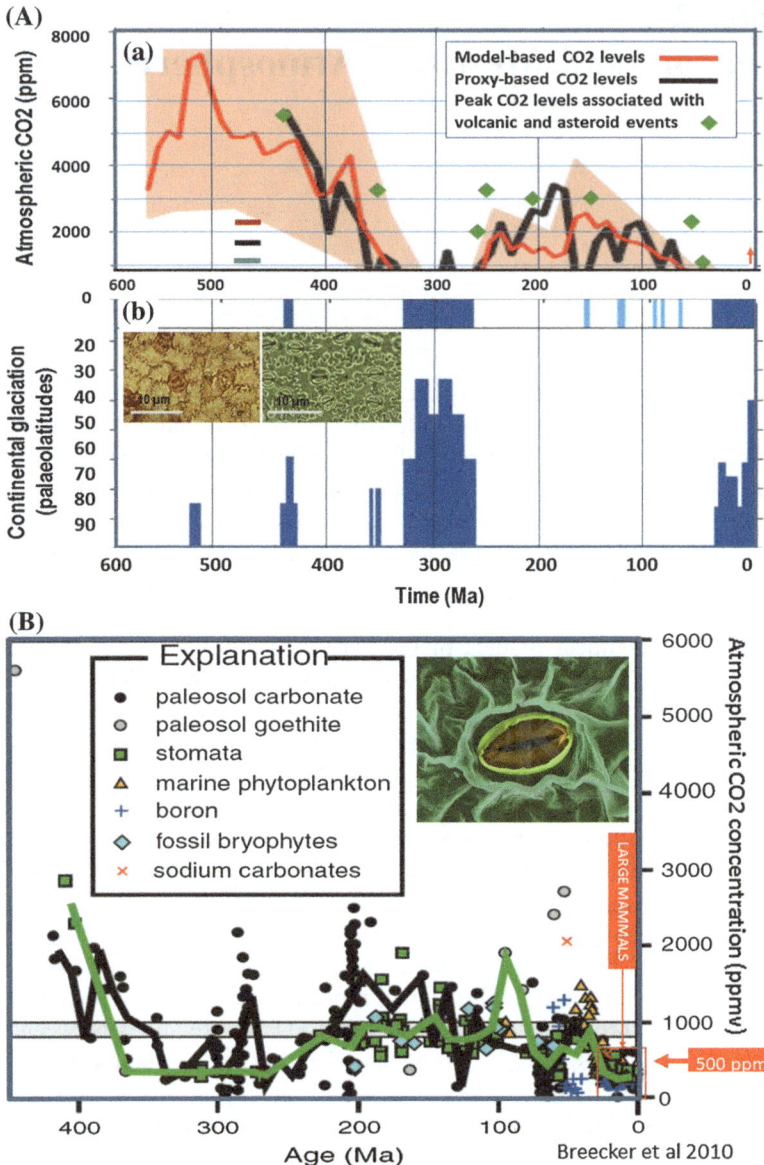

Fig. 2.1 **A** (**a**) Evolution of CO_2 and occurrence of ice ages and cold phases from the Cambrian (542 Ma) to the present. Ice ages from proxy-based temperatures and sedimentological data (Royer et al. 2004, Fig. 2; geological society of America, by permission); **b** Photomicrographs of stomata pores: *left*—fossil leaf cuticle of the fern affinity Stenochlaena from just after the Cretaceous/Tertiary (K/T) boundary. *Right*—the fern's nearest living relative, Stenochlaena palustris. The stomatal index of the fossil cuticle is considerably lower than the extant cuticle, indicating that CO_2 was higher directly after the K/T boundary than today (Royer 2008, Fig. 2; geological society of America, by permission). *Green diamonds*: CO_2 levels corresponding to asteroid impacts, volcanic and mass extinction events. **B** Compilation of Phanerozoic atmospheric CO_2 based on different proxies (Breecker et al. 2009, Fig. 2; proceedings of the national academy of science, by permission)

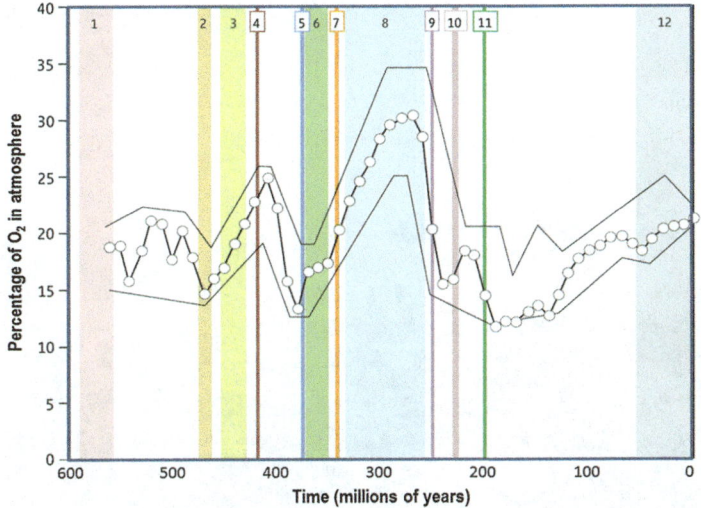

Fig. 2.2 Oxygen variations through the Phanerozoic. The *upper* and *lower* boundaries are estimates of error in modeling atmospheric O_2 concentration. The numbered intervals denote important evolutionary events that may be linked to changes in O_2 concentration (Berner et al. 2007, Fig. 1; *courtesy* R.A. Berner; American association for advancement of science and Elsevier, by permission)

leads to acidification, retarding phytoplankton oxygen production, while on land droughts and fires reduce the carbon/oxygen fractionation effects of plant photosynthesis.

A schematic illustration of the carbon cycle is presented in Fig. 2.3. The evolution of life through the Phanerozoic era can be described in three parts: Palaeozoic (early life $\sim$542–251), Mesozoic (Middle life $\sim$251–65 Ma) and Cenozoic (new life 65–0 Ma), separated by abrupt extinctions at $\sim$251 Ma and 65 Ma (Tables A1 and A2), including the following major mass extinctions of genera (after Keller 2005; Fig. 4.1):

1. End-Early Cambrian—513 ± 2 Ma $\sim$42 % extinction of genera;
2. End-Ordovician—443 ± 7 Ma—$\sim$57 % extinction of genera;
3. Late and end Devonian—371–349 Ma $\sim$58 and 30 % extinction of genera;
4. Permian–Triassic boundary—251 ± 0.4 Ma $\sim$80 % extinction of genera;
5. Late Triassic (Norian)—216 Ma $\sim$34 % extinction of genera;
6. K-T boundary—65.5 ± 0.3 Ma $\sim$46 % extinction of genera.

During the Cambrian to the late Devonian ($\sim$542–370) the dominance of high atmospheric CO_2 levels was interrupted by relatively brief cold periods and ice ages, including the late Ordovician glaciation ($\sim$446–443 Ma), late and end Devonian to early Carboniferous glaciations (371–349 Ma—$\sim$58 and 30 % extinction of genera) and Carboniferous-Permian glaciation (326–267 Ma) (Fig. 2.1). Intermittent rises in ocean temperatures with consequent depletion in

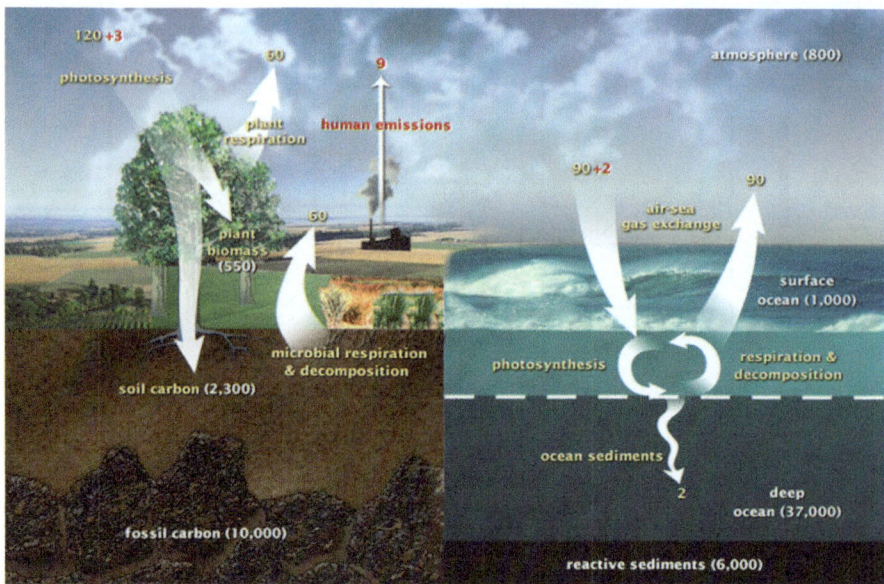

Fig. 2.3 The carbon cycle (values in GtC—billion tons carbon). The diagram portrays the fast carbon cycle shows the movement of carbon between land, atmosphere, and oceans. *Yellow numbers* are natural fluxes, and *red* are human contributions in gigaton of carbon per year. *White numbers* indicate stored carbon. (U.S. DOE, biological and environmental research information system, NASA earth observatory. http://earthobservatory.nasa.gov/Features/CarbonCycle/)

oxygen led to anoxia events, a Mesozoic example being a ∼94 Ma anoxia event represented by increase in burial of organic carbon (Barclay et al. 2010). These authors report a rise in CO_2 from 370 to 500 ppm consequent on volcanic eruptions, followed by a decrease of ∼26 % associated with sequestration of marine organic carbon.

Based on proxy and model estimates of Phanerozoic carbon and oxygen levels (Beerling et al. 2001; Berner 2004, 2005, 2006, 2009; Berner et al. 2007; Beerling and Royer 2011) mean CO_2 levels during the early Paleozoic (∼542–370 Ma) varied in the range of ∼5,000–2,000 ppm (Fig. 2.1). The models are based on fractionation of carbon isotopes between carbonate and organic matter in sediments, and different rates of weathering of basalts and granitic rocks, using a GEOCARBSULF model. Berner et al. (2007) point to several stages in the evolution of oxygen related to biological evolution (Fig. 2.2):

1. From ∼540 Ma proliferation of marine animals associated with rise of oxygen to ∼15–20 % O_2;
2. The rise of land plants at the end-Silurian (∼420 Ma) when O_2 reached ∼25 %;
3. Major increase in O_2 to levels of ∼25–35 % through the Carboniferous-Permian (∼340–220 Ma), when gigantism developed among arthropods and reptiles thank to the increase in oxygen;

4. A rise in O_2 from ~ 15 to ~ 20 % from the end-Triassic ~ 201 Ma through to the Cenozoic, when the size of dinosaurs and subsequently mammals was linked to oxygen levels.
5. Climate models based on mid-Cenozoic conditions suggest a glaciation threshold of ~ 500–600 ppm CO_2 according to Zachos et al. (2001) and 560–1,120 ppm according to Pollard and DeConto (2005).

Based on the study of temperature and biogenic proxies and on carbon cycle models, Cambrian CO_2 levels were as high as 2,000–4,000 ppm and Silurian levels about 3,000 ppm (Fig. 2.1c). Following Royer (2006) main stages in the Phanerozoic atmospheric history include:

Late Ordovician glaciation (445.6–443.7 Ma): Widespread glaciations at the Ordovician–Silurian boundary (Frakes et al. 1992) persisted into the early Silurian. CO_2 data are sparse and one point suggests $\sim 5,600$ ppm, whereas Berner's GEOCARB III model (Berner 2006) suggests $\sim 4,200$ ppm. However, the time span of these CO_2 levels is not clear. According to Kump et al. (1999) CO_2 levels declined during this transition from 5,000 to 3,000 ppm. Analysis of radiative forcing indicates that, if the CO_2-Antarctic ice threshold for the present-day Earth is 500 ppm, in the Late Ordovician insolation of about 4 % below the present renders a CO_2 glaciation threshold as high as 2,240–3,920 ppm (Royer 2006; Kump et al. 1999). These estimates remain to be confirmed by the multiple proxy studies.

Late Devonian-early Carboniferous glaciations (371–349 Ma): Two brief but widespread glaciations analogous to the Permo-Carboniferous glaciation and associated with the position of Gondwana at high latitudes have been identified (Royer 2006). There is evidence for ice near the Frasnian–Fammenian boundary (374.5 Ma) when CO_2 levels dropped sharply from 3,300 to 350 ppm, but the main glaciation occurred in the Fammenian (361.4–360.6 Ma). Early Carboniferous glaciation appears to span the 353–349 Ma period. CO_2 level of $\sim 1,000$ ppm was measured for 3.10^6 years before the Late Devonian glaciation and 1,300 ppm for 7.10^6 years following the early Carboniferous glaciation (Royer 2006).

Carboniferous-Permian glaciation (326–267 Ma): This glaciation constitutes the longest and most extensive Phanerozoic glaciation, during which O_2 levels declined to near or less than 500 ppm. Calibrated CO_2 values are based on arborsecent lycopsids from equatorial Carboniferous and Permian swamp communities, obtaining concentrations of 344 ppm and 313 ppm for the late Carboniferous and early Permian, respectively. The position of Australia and Antarctica near the South Pole (Frakes et al. 1992; Beerling 2002a) resulted in extensive glaciation over a period of near 60 million years (Crowley 1999; Crowley and Berner 2001; Royer et al. 2004). The low CO_2 levels are in agreement with glaciological evidence for the presence of continental ice and coupled models of climate and ice-sheet growth on Pangaea. A period of 1,500 ppm CO_2 occurs at 300 Ma, defining two glacial periods separated by an ice-free phase at 311.7–302 Ma (Isbell et al. 2003). CO_2 levels are consistently low during the first glacial phase (326.4–311.7 Ma), with most data <500 ppm, folloed by rise to ∼1,500 ppm at ∼300 Ma, then drop sharply to below 500 ppm. After 267 Ma levels CO_2 levels rise to ∼1,000 ppm.

Early Jurassic to Cretaceous cool phases (184–66.5 Ma): Marked climate fluctuations occurred during 184–183 Ma. Tropical conditions dominated much of the Mesozoic and little evidence exists for permanent ice during this period (Frakes et al. 1992; Eyles 1993; Price 1999; Zachos et al. 2001), consistent with CO_2 levels in the order of ∼1,000 ppm and higher (Fig. 2.1). However, these conditions were interrupted by cooler climates, such as in the mid-Jurassic to early Cretaceous (171.6–106 Ma) (Frakes et al. 1992) and multiple short cooling events between 184 and 66.5 (Royer 2006).

References

Barclay RS, McElwain JC, Sageman B (2010) Carbon sequestration activated by a volcanic CO_2 pulse during ocean anoxic event. Nat Geosci 3:205–208

Beerling DJ (2002a) CO_2 and the end-triassic mass extinction. Nature 415:386–387

Beerling DJ, Royer D (2011) Convergent cenozoic CO_2 history. Nat Geosci 4:418–420

Beerling DJ, Osborne CP, Chaloner WG (2001) Evolution of leaf-form in land plants linked to atmospheric CO2 decline in the Late Palaeozoic era. Nature 410:352–354

Berner RA (2004) The phanerozoic carbon cycle: CO_2 and O_2. Oxford University Press, New York

Berner RA (2005) The carbon and sulfur cycles and atmospheric oxygen from middle Permian to middle Triassic. Geochim Cosmochim Acta 69:3211–3217

Berner RA (2006) GEOCARBSULF: a combined model for phanerozoic atmospheric O_2 and CO_2. Geochim et Cosmochim Acta 70(23):5653–5664

Berner RA (2009) Phanerozoic atmospheric oxygen new results using the GEOCARBSULF model. Am J Sci 309:603–606

Berner RA, Vanderbrook JM, Ward PD (2007) Oxygen and evolution. Science 316:557–558

Crowley JC (1999) Pre-Mesozoic ice ages: their bearing on understanding the climate system. Geol Soc Am Mem 192. Boulder

Breecker DO, Sharp ZD, McFadden LD (2009) Atmospheric CO_2 concentrations during ancient greenhouse climates were similar to those predicted for A.D. 2100. Proc Nat Acad Sci 107:576–580

Crowley TJ, Berner RA (2001) CO_2 and climate change. Science 292:870–872

Eyles N (1993) Earth's glacial record and its tectonic setting. Earth Sci Rev 35:1–248

Frakes LA, Francis JE, Syktus JI (1992) Climate modes of the Phanerozoic. Cambridge University Press, Cambridge

Isbell JL, Miller MF, Wolfe KL, Lenaker PA (2003) Timing of late paleozoic glaciation in Gondwana: was glaciation responsible for the development of Northern hemisphere cyclothems? In: Chan MA, Archer AW (eds) Extreme depositional environments: Mega end members in geologic time. Geol Soc Am Spec Pap 340, Boulder, p 5–24

Keller G (2005) Impacts volcanism and mass extinction: random coincidence or cause and effect? Aust J Earth Sci 52:725–757

Kump LR, Arthur MA, Patzkowsky ME, Gibbs MT, Pinkus DS, Sheenan PM (1999) A weathering hypothesis for glaciation at high atmospheric pCO_2 during the late Ordovician. Palaeoclimatol Palaeogeogr Palaeoecol 152:173–187

Pollard D, DeConto RM (2005) Hysteresis in cenozoic antarctic ice sheet variations. Glob Planet Change 45:9–21
Price GD (1999) The evidence and implications of polar ice during the mesozoic. Earth Sci Rev 48:183–210
Royer DL (2006) CO_2-forced climate thresholds during the phanerozoic. Geochim Cosmochim Acta 70:5665–5675
Royer DL (2008) Linkages between CO_2, climate, and evolution in deep time. Proc Nat Acad Sci 105:407–408
Royer DL, Berner RA, Montañez I, Neil P, Tabor J, Beerling DJ (2004) CO_2 as a primary driver of phanerozoic climate. GSA Today 14:3
Strauss H, Peters-Kottig W (2003) The paleozoic to mesozoic carbon cycle revisited: the carbon isotopic composition of terrestrial organic matter. Geochem Geophys Geosystems 4(10)
Zachos J, Pagani M, Sloan L, Thomas E, Billups K (2001) Trends, rhythms, and aberrations in global climate 65 Ma to present. Science 292:686–693

Chapter 3
Cenozoic Atmospheres and Early Hominins

Abstract The Cenozoic era includes four components (A) post K-T impact warming culminating with the Paleocene-Eocene hyperthermal at ~55 Ma; (B) long term cooling ending with a sharp temperature plunge toward formation of the Antarctic ice sheet from 32 Ma; (C) a post-32 Ma era dominated by the Antarctic ice sheet, including limited thermal rises in the end-Oligocene, mid-Miocene and end-Pliocene, and (D) Pleistocene glacial-interglacial cycles. Hominin evolution in Africa occurred during a transition from tropical to dry climates punctuated by alternating periods of extreme orbital forcing-induced glacial-interglacial cycles, suggesting variability selection of Hominids.

The onset of the Cenozoic, marked by the ~65 Ma K-T asteroid impact event (Alvarez et al. 1980) (Fig. 4.7), which occurred at a time of low global atmospheric CO_2 concentrations (Figs. 2.1, 4.6), resulted in an increase in atmospheric CO_2 from ~400 to ~2,400 ppm (Beerling et al. 2002), subsequently decreasing to ~300 ppm within ~500 kyr. A succeeding warming trend culminated in the Paleocene-Eocene thermal maximum (PETM) at ~55.9 Ma, which involved a release of some ~2,000 to 3,000 billion ton carbon (GtC) as methane, elevating atmospheric CO_2 to near ~2,000 ppm at a rate of approximately ~0.13 ppm/year and mean temperatures rise of several degrees Celsius (Zachos et al. 2008; Cui et al. 2011) (Figs. 3.2, 3.3). Elevated atmospheric carbon led to ocean acidification from ~8.2 to ~7.5 pH and to an extinction of 35–50 % of benthic foraminifera over the course of ~1,000 years (Zachos et al. 2008). Other consequences included a global expansion of subtropical dinoflagellate plankton, appearance of modern orders of mammals (including primates), a transient dwarfing of mammalian species, and migration of large mammals from Asia to North America.

Hominids—the group consisting of all modern and extinct Great Apes (that is, modern humans, chimpanzees, gorillas and orang-utans plus all their immediate ancestors). Hominin—the group consisting of modern humans, extinct human species and all our immediate ancestors (including members of the genera Homo, Australopithecus, Paranthropus and Ardipithecus).

A. Y. Glikson, *Evolution of the Atmosphere, Fire and the Anthropocene Climate Event Horizon*, SpringerBriefs in Earth Sciences, DOI: 10.1007/978-94-007-7332-5_3, © The Author(s) 2014

In the wake of the PETM event atmospheric greenhouse levels progressively cooled to about ∼34 Ma (Frakes et al. 1992; Eyles 1993; Zachos et al. 1992, 2001; Ruddiman 1997, 2008; Zachos et al. 2001; Royer et al. 2004; Royer 2006). The cooling is consistent with the CO_2 record, indicating a decline from high CO_2 levels of 1,200 ppm at 44 Ma to <500 ppm before the onset of the first cool event at 42 Ma (Royer 2006). Evidence from sea level, temperature, and calcite compensation depth suggests brief glaciations during the upper Eocene at about ∼42 to 41, ∼39 to 38, and ∼36.5 to 36 Ma (Browning et al. 1996). The gradual decline in mean global CO_2 and temperatures ended at ∼34 Ma with a sharp drop of atmospheric CO_2 to below 500 ppm, a decline of mean global temperature by ∼5 °C and the onset of the Antarctic ice sheet (Zachos et al. 2001, 2008) (Fig. 3.1). Following ∼34 Ma CO_2 levels have been mainly below ∼500 ppm, an exception being a late Oligocene (∼25 Ma) warming (Zachos et al. 2001) when CO_2 levels higher than 500 ppm are recorded. The upper Eocene temperature decline is interpreted in terms of CO_2 capture associated with erosion of the rising Himalayan and Alpine mountain chains (Ruddiman 1997). The sharp freeze at 34 Ma (Figs. 3.1, 3.2, 3.4) is likely related to the opening of the Drake Passage between South America and west Antarctica and onset of the circum-Antarctic current, isolating Antarctica from the influence of warmer currents.

The sharp decline in temperature at ∼34 Ma was preceded by a large cluster of ∼35.7 to 35.5 Ma asteroids (Popigai, 100 km, 35.7 ± 0.2 Ma; Chesapeake Bay, 85 km, 35.5 ± 0.3 Ma; Mount Ashmore, >50 km, E-O Boundary) (Glikson et al. 2010), which may have been a factor in enhancing the tectonic breakdown of the previous Antarctic-South America land link.

The formation of the Antarctic ice sheet between 34 and 33.5 Ma was associated with a sharp decline of atmospheric CO_2 to below 750 ppm (Fig. 3.4). Studies of carbonate microfossils using the $\delta^{11}B$ proxy allow tracing of CO_2 levels through the Eocene–Oligocene transition in Tanzania (Pearson et al. 2009), indicating the decline in CO_2 commenced before inception of the Antarctic ice sheet, followed by recovery between 33.5 and 33.3 Ma to levels as high as ∼1,100 ppm, succeeded by a decline of CO_2 levels to about or below ∼500 ppm about 30 Ma (Fig. 3.4). Once the Antarctic ice sheet formed hysteresis allowed its persistence despite transient rise in the GHG. However, no $\delta^{18}O$ evidence for a rise in temperatures is observed during the CO_2 rebound period. This is in part explained by major reduction in sea surface area and thereby reduced CO_2 sequestration rates. Other possible factors were release of organic carbon from geological reservoirs, changes in ocean productivity and circulation patterns, and variations in rock weathering rates (Pearson et al. 2009).

Following the isolation of Antarctic ice sheet the global climate was affected by major feedback processes, including

1. Albedo enhancement by continental ice sheets, sea ice and mountain glaciers (Hansen et al. 2007, 2008; Overpeck et al. 2006);
2. Cold currents and weather fronts emanating from the Antarctic vortex and the circum-Antarctic current;

3 Cenozoic Atmospheres and Early Hominins

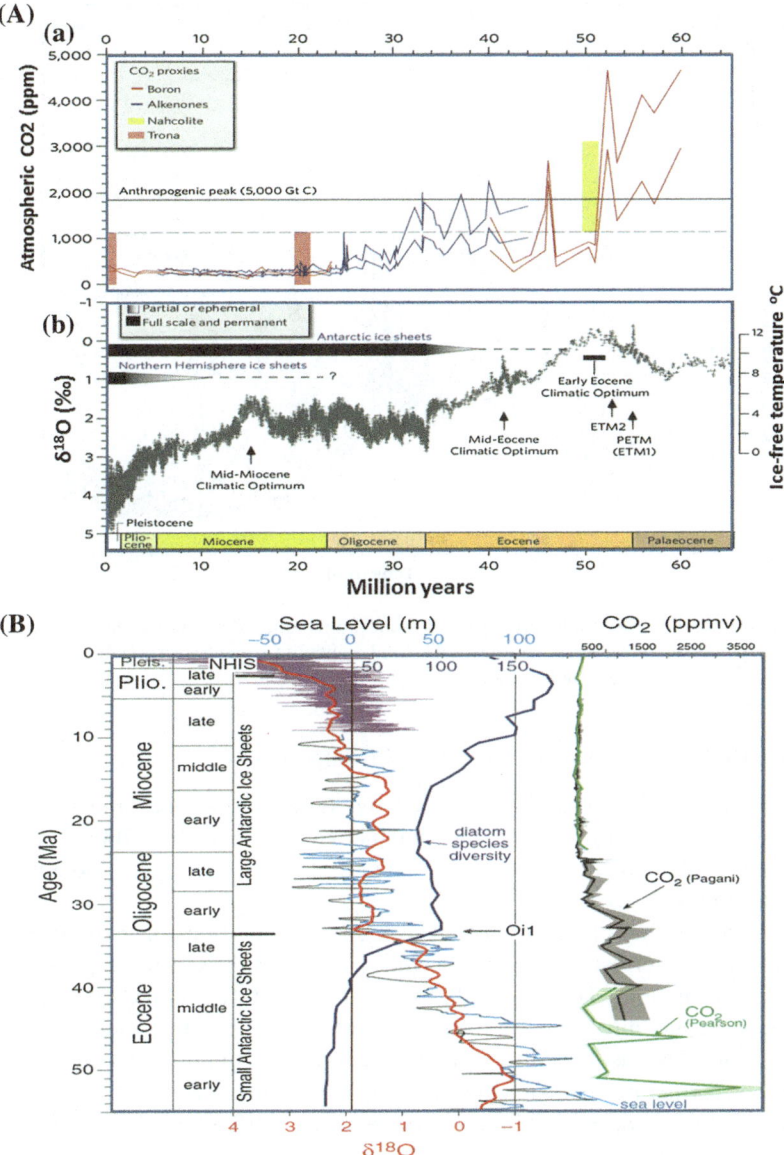

Fig. 3.1 A Cainozoic CO_2 and temperature trends showing (**a**) CO_2 levels based on boron, alkenones, nahcolite and trona proxies; **b** development of the Antarctic ice sheet and northern hemisphere glaciations, ice-free water temperatures and $\delta^{18}O$ temperature proxies (Zachos et al. 2008, Fig. 2; nature, by permission). Note the PETM event (Fig. 3.3) and the end-Eocene freeze event (Figs. 3.4). **B** Comparison of the global sea level (blue indicates intervals constrained by data; black indicates estimated low-stands, benthic foraminiferal $\delta^{18}O$ values, diatom diversity curve and atmospheric CO_2 estimates derived from alkenones and boron isotopes. NHIS—Northern Hemisphere ice sheets. The benthic foraminiferal curve is the Atlantic synthesis which is smoothed to remove periods shorter than 1 m.y. (Miller et al. 2009, Fig. 1; the geological society of America, by permission)

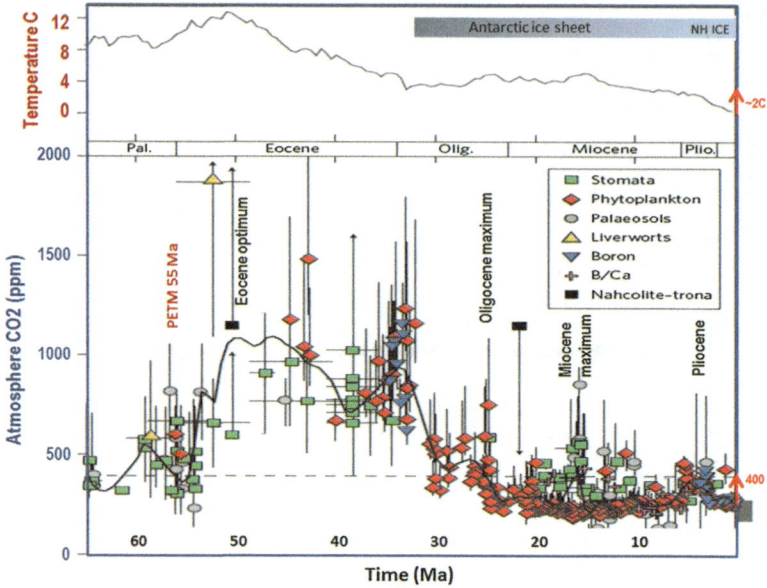

Fig. 3.2 Cainozoic CO_2 multi-proxy-based trends, including data based on stomata, phytoplankton, paleosols, liverworts, boron, B/Ca and nahcolite-trona (Beerling and Royer 2011, Fig. 1; Nature Geoscience, by permission)

3. Enhanced sequestration of CO_2 by cold water.

As atmospheric CO_2 levels declined the Earth surface became increasingly exposed to orbital forcing of the Milankovic cycles (Roe 2006) (Fig. 3.5), including weak solar cycles such as the 11-years sun spot cycle (Solanki 2002; Bard and Frank 2006).

Once the thermal blanketing effect of high concentrations of greenhouse gases is removed, the terrestrial climate becomes increasingly sensitive to minor variations in insolation. The molecular resonance of GHG (water vapor, carbon dioxide, methane, nitric oxides, ozone), absorbing and emitting thermal radiation, can alternately act as a driver of global temperatures and/or as feedback to insolation. The former mechanism controlled the Pleistocene glacial-interglacial cycles (Figs. 3.6b, 3.7, 3.8, 3.9), while an independent GHG effect is manifest at the PETM (Fig. 3.3).

Eocene cooling and the following ice ages saw the flourishing of land mammals, which in the Mesozoic were mostly limited to small burrowing species. The overall cooling was interrupted by a number of warming events which triggered sea level rises, including the late Oligocene ($\sim$25 Ma), early and mid-Miocene (22 Ma, 17–14 Ma), mid-Pliocene (3.1–2.9 Ma) and Pleistocene (2.8 Ma–10 kyr) interglacial peaks (Fig. 3.6b, 3.7, 3.8, 3.9). These oscillations, generated by Milankovic orbital cycles, including periodic changes in the eccentricity, axial obliquity and precession of the Earth relative to the sun (Fig. 3.5), display

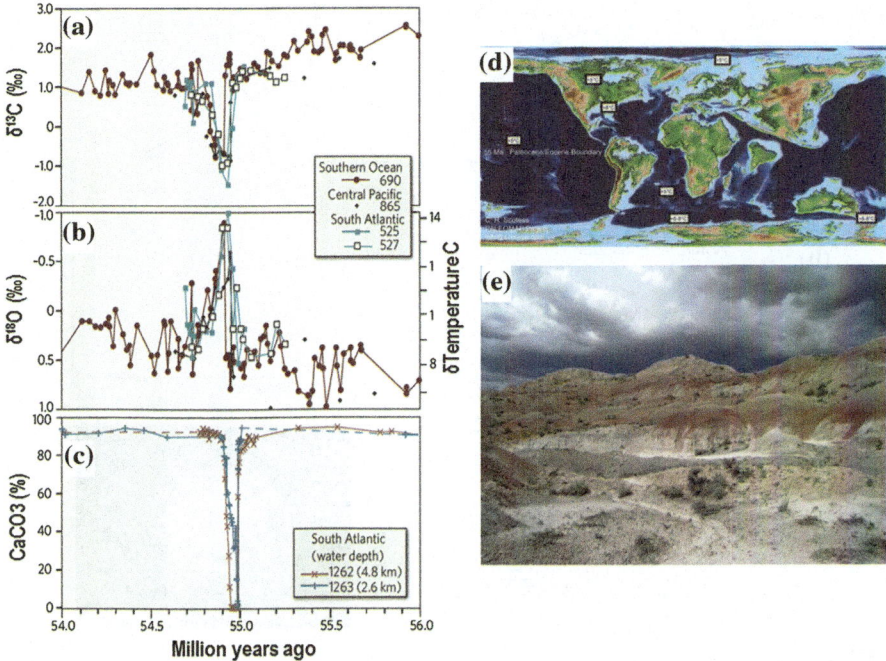

Fig. 3.3 The Paleocene-Eocene thermal maximum (*PETM*) represented by sediments in the Southern Ocean, central Pacific and South Atlantic oceans. The data indicate **a** deposition of an organic matter-rich layer consequent on extinction of marine organisms; **b** lowering of $\delta^{18}O$ values representing an increase in sea water temperature, and **c** a sharp decline in carbonate contents of sediments representing a decrease in pH (increase in acidity) (Zachos et al. 2008, Fig. 3; nature, by permission); **d** PETM map—*courtesy* Appy Sluijs; **e** photograph showing the Willwood Formation, Bighorn Basin, Wyoming, representing the final stages and recovery phase of the Paleocene-Eocene thermal maximum (*PETM*) (courtesy Cesca McInerney)

increasingly sharper amplitudes with time (Fig. 3.6b). Peak Pliocene conditions, reached when CO$_2$ levels were about <400 ppm, saw temperatures about 2–3 °C higher and sea level 25 ± 12 m above late Holocene pre-industrial levels (Chandler et al. 2008). Sea levels continued to fluctuate through the Pleistocene in direct relation to temperatures.

About 4.0 Ma, following a rise of the Panama cordillera and closing of the Panama straits, separation of the Pacific and Atlantic Ocean basins augmented longitudinal ocean circulation, enhancing the Pacific Gyre, the El-Nino Southern Oscillation (ENSO) cycle (Fig. 3.6b) and the North Atlantic Thermohaline Current (Fig. 3.10). The consequent intensification of cross-latitude circulation patterns enhanced the cold Humboldt Current along the South America west coast, reaching low latitudes, and associated development of the La Nina phase of the ENSO in the central Pacific Ocean (Fig. 3.6b) (Fedorov et al. 2006). Pliocene orbital forcing cycles, mostly modulated by the 19–23 kyr-long precession-controlled cycles are represented by organic-rich sapropel layers (Fig. 3.7). From

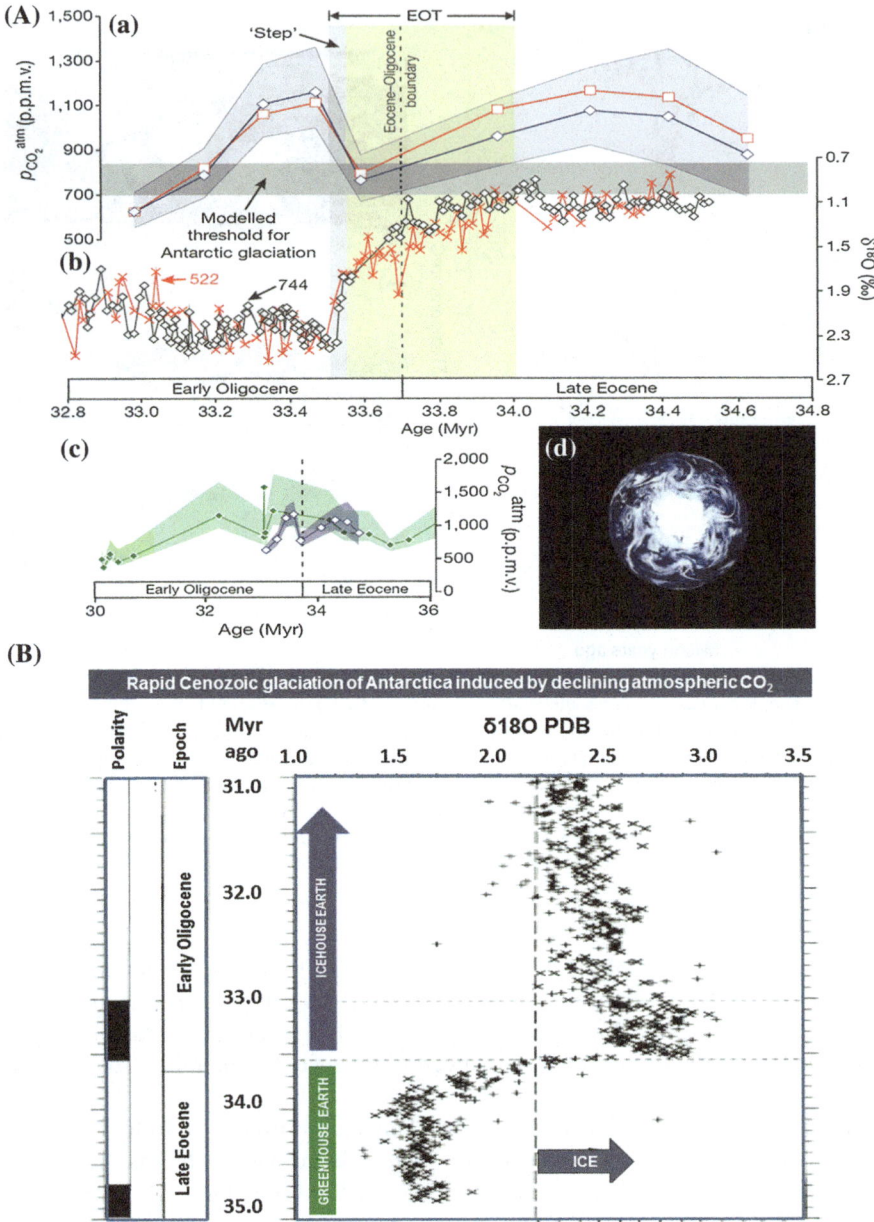

Fig. 3.4 The end-Eocene freeze: a Reconstructed atmospheric CO_2 levels from Boron isotopes (^{11}B); b deep-sea oxygen isotopes ($\delta^{18}O$) are from DSDP site 522 (*red crosses*) and ODP site 744 (*black diamonds*); c Alkenones proxy CO_2 estimates. The grey band is the threshold for Antarctic glaciation; Note the sharp drop in CO2 from ~1,100 to ~700 ppm and of temperatures (expressed by rising $\delta^{18}O$) to ~33.6, followed by a transient rise in CO_2 (Pearson et al. 2009, Fig. 3; Nature, by permission); d Satellite image of Antarctica (*NASA*). B Rapid cenozoic glaciation of Antarctica induced by declining atmospheric CO_2 (Pollard and DeConto 2005, Fig. 2c; Nature, by permission)

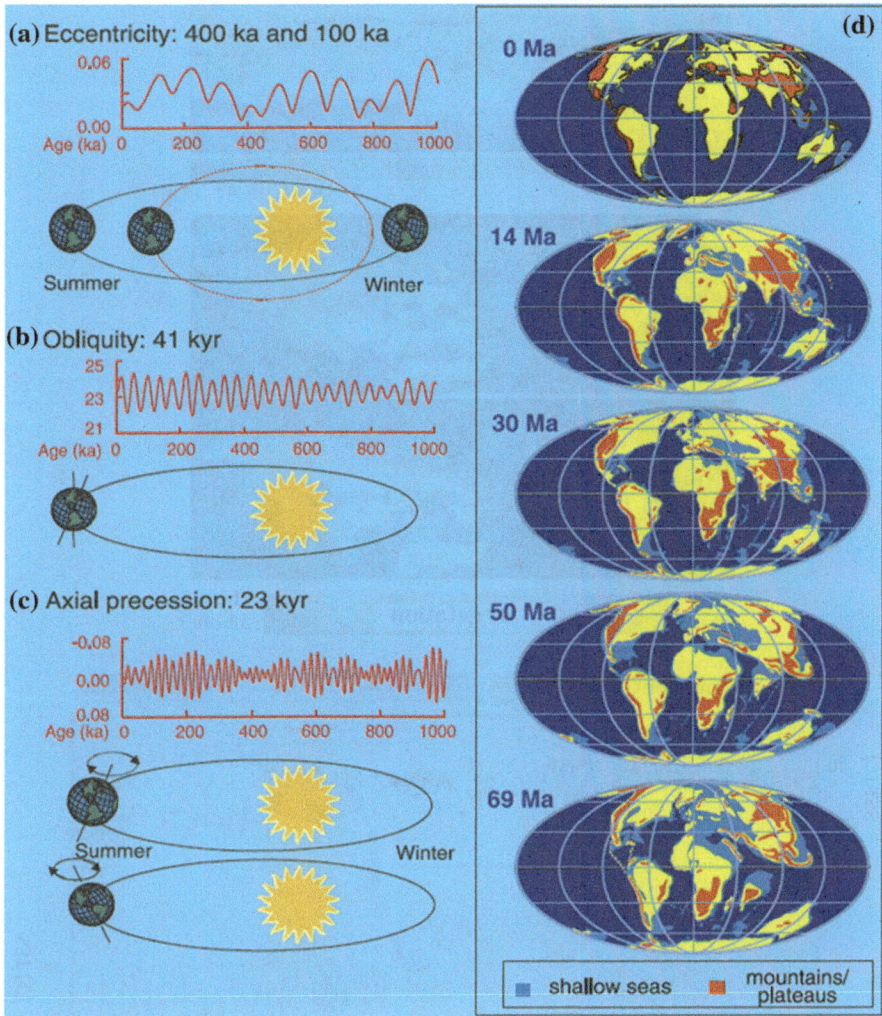

Fig. 3.5 Orbital components of Milankovic cycles and Cenozoic paleogeography. Orbital components include eccentricity (400 and 100 kyr), obliquity (41 kyr), and precession (23 and 19 kyr). **a** Eccentricity refers to the shape of Earth's orbit around the sun; **b** obliquity refers to the tilt of Earth's axis relative to the plane of the ecliptic varying between 22.1° and 24.5°. A high angle of tilt increases the seasonal contrast, most effectively at high latitudes, where winters will be colder and summers hotter as obliquity increases; **c** Precession is the wobble of the axis of rotation describing a circle in space with a period of 26 kyr. The periods of the precession modulated by eccentricity of 23 and 19 kyr are observed in geological records; **d** paleogeographic maps back 69 Ma. (Zachos et al. 2001, Fig. 1; Courtesy J.C. Zachos; American Association for the Advancement of Science, by permission)

36   3 Cenozoic Atmospheres and Early Hominins

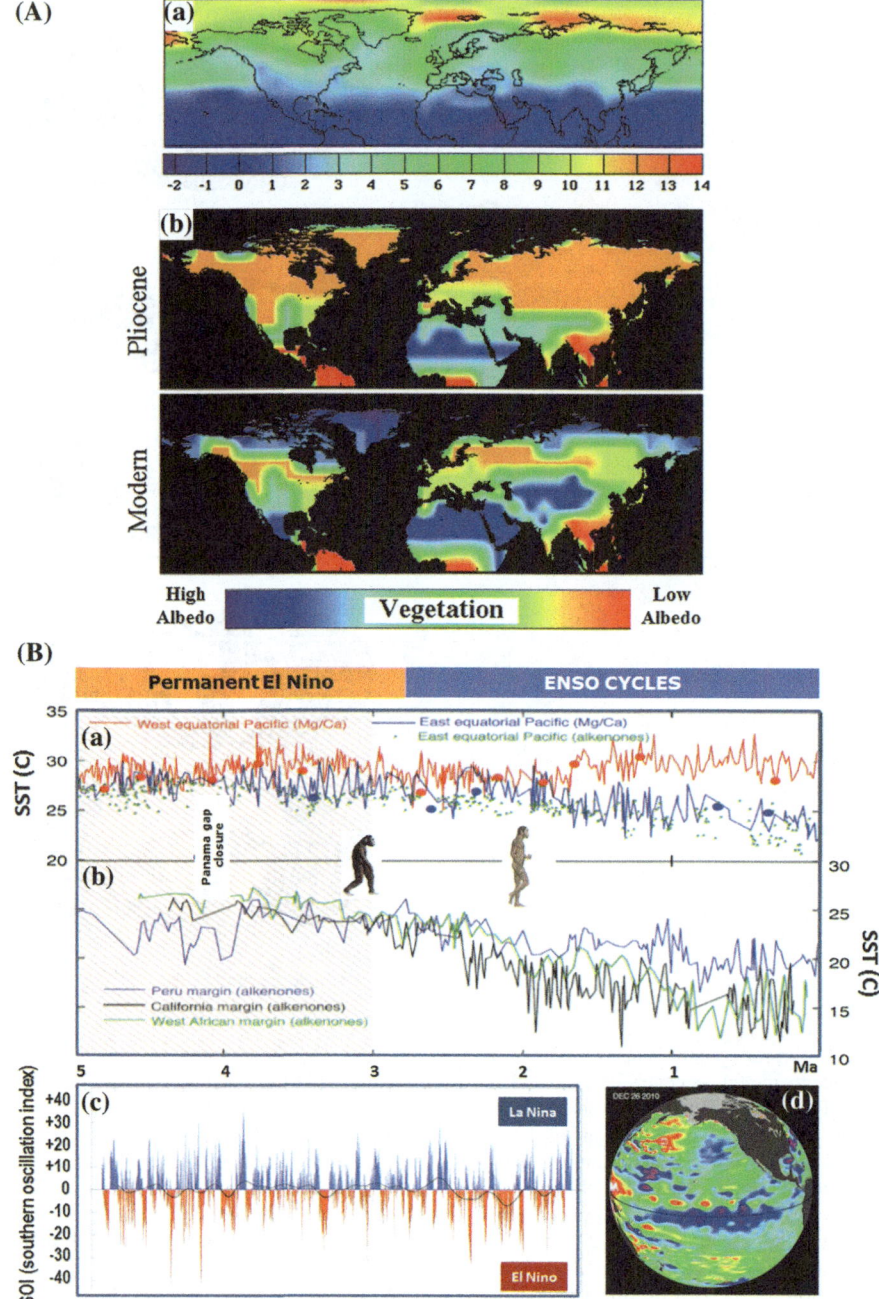

Fig. 3.6 A Model and paleo-data based comparisons between Pliocene and late Holocene climates. **a** Pliocene Northern Hemisphere surface air temperatures relative to Holocene pre-industrial temperatures. Model and data indicate significantly warmer temperatures at high latitudes and diminished relative warming nearer to the equator; **b** Pliocene and modern vegetation albedo distribution (NASA—National Aeronautics and Space Administration/ Goddard Institute for Space Studies; http://www.giss.nasa.gov/research/features/ 199704_pliocene/page2.html. **B** Evolution of the ENSO displaying overall cooling since ∼3 Ma; **a** Sea surface temperatures (SST) based on Mg/Ca and alkenones paleo-temperature proxies. The plot displays near coincidence between East Pacific and West Pacific sea surface temperatures before about 4 Ma, incipient divergence between cooler East Pacific and warmer West Pacific temperatures between about 3 and 2 Ma and increasingly marked divergence in temperatures from about 2 Ma when a strong ENSO polarity ensued; **b** Alkenones proxy evidence for cooling of the Humboldt current (Peru margin), California current and West African (Namibia) current parallel to the development of ENSO polarity (B and C from Fedorov et al. 2006, Fig. 3; American Association for Advancement of Science, by permission); **c** ENSO oscillations during 1880–2012, displaying intensification of El-Nino events about 1900 and between 1980 and 2000 (NASA); **d** A La Nina event in 2010 (NASA; http://www.ncdc.noaa.gov/ teleconnections/enso/enso-sigevents.php)

about ∼2.8 Ma orbital forcing was dominated by 41 kyr-long obliquity-modulated cycles associated with increased generation of dust related to drying and glacial erosion and winds. From about ∼1.8 Ma increased amplitude of ∼41 kyr-long cycles was associated with marked increased in the ENSO polarity (Fig. 3.6b) and the frequency of the La-Nina phase. From about ∼0.9 Ma ∼100 kyr-long eccentricity-dominated cycles became dominant, displaying glacial-interglacial temperature variations of up to 6 °C and glacial dust levels in marine sediments of up to 40 % (Petit et al. 1999; EPICA 2004). The increased polarity of glacial-interglacial cycles and the abrupt nature of the glacial terminations are attributed to the decrease in atmospheric CO_2 levels, allowing stronger orbital-driven thermal pulsations, and to triggering by solar maxima of amplifying feedbacks, including ice/melt albedo flip and CO_2 release from warming seas and drying/burning plants (Hansen et al. 2007, 2008).

The evolution of hominins in the rift valley region of Eastern Africa (Figs. 3.11, 3.12) has been intimately related to climatic variability controlled by a combination of tectonic changes and insolation cycles (Maslin et al. 2001; Maslin and Christensen 2007; Maslin and Trauth 2009; Trauth et al. 2007, 2010). Down-faulting of rift valleys during 5.5–3.7 Ma resulted in formation of lakes and sheltered environments for the evolution of hominins during the retreat of rain-forests and opening of savanna, with further rifting about 1.2 Ma. Deep lakes developed during ∼3.20 to 2.95, ∼3.4 to 3.3, 4.0 to 3.9, ∼4.7 to 4.3, 2.7–2.5, 1.9–1.7 and 1.1–0.9 Ma, periods, which correspond to low-dust relatively humid periods in North Africa represented by sapropel (Fig. 3.7). The 2.7–2.5 Ma Lake phase corresponds to intensification of the Northern Hemisphere Glaciation, the 1.9–1.7 Ma Lake phase to development of the Walker Circulation, and the 1.1–0.9 Ma Lake phase to initiation of the Mid-Pleistocene transition from 41 kyr cycles to 100 kyr cycles (Berger and Jansen 1994). Moisture levels and the filling of the lakes were sensitively related to movements of the Intertropical

Fig. 3.7 An outcrop of late Miocene (9.3–8.4 Ma) sediments displaying dark organic matter (sapropel)-rich cycles representing humid periods, Gibliscemi, Sicily. The sapropel units represent 19–23 kyr-long precession-induced monsoonal periods affecting increased organic matter production transported through the river Nile to the Mediterranean sea. Precession cycles are grouped in sets of four, representing ∼100 kyr-long eccentricity cycles (deMenocal 2004, Fig. 6; Elsevier, by permission)

Convergence Zone (ITCZ) and the Indian Ocean monsoon (Maslin and Trauth 2009). Major lacustrine episodes during 2.7–2.55 Ma consist of five paleo-lake phases which correspond to the precession cycle of 23 kyr and extreme climate variability (Deino et al. 2006). Carbon isotopes from both soil carbonates and biomarkers from deep-sea sediments indicate progressive vegetation shift from C3 (trees and shrubs) to C4 (tropical grasses) plants during the Plio-Pleistocene (Feakins et al. 2005), reflecting increased aridity. Extreme climate variability is supported by paleo-diet reconstructions of mammals and hominins (Teaford and Ungar 2000).

The close relations between orbital forcing, climate variability and vegetation in East Africa resulted in the rapid appearance and disappearance of ephemeral lakes over periods of hundreds to thousands of years, with implications for the speciation and dispersal of mammals (Deino et al. 2006). Lake formation peaks during periods of high eccentricity appear to have near-coincided with the appearance of Olduvan stone culture about ∼2.6 Ma and Acheulean stone culture about 1.9–1.8 Ma (Figs. 3.11, 6.2). 12 out of the 15 hominin species first appeared during alternating wet-dry periods (Maslin and Trauth 2009), including *Homo habilis*, *H. rudolfensis*, *H. erectus* and/or *H. ergaster* (Figs. 3.11, 6.2). Transitions between different climate states may have been gradual, abrupt or extreme, the latter supporting variability selection in human evolution (Potts 1998). Extreme variability is documented at Lake Baringo where diatomite units and fish-bearing units grade into high-energy terrestrial facies (Maslin and Trauth 2009). Variability selection invokes key hominin adaptations during periods of rapid change, including development of bipedal locomotion, high brain/body mass relations and complex human social behavior.

3 Cenozoic Atmospheres and Early Hominins

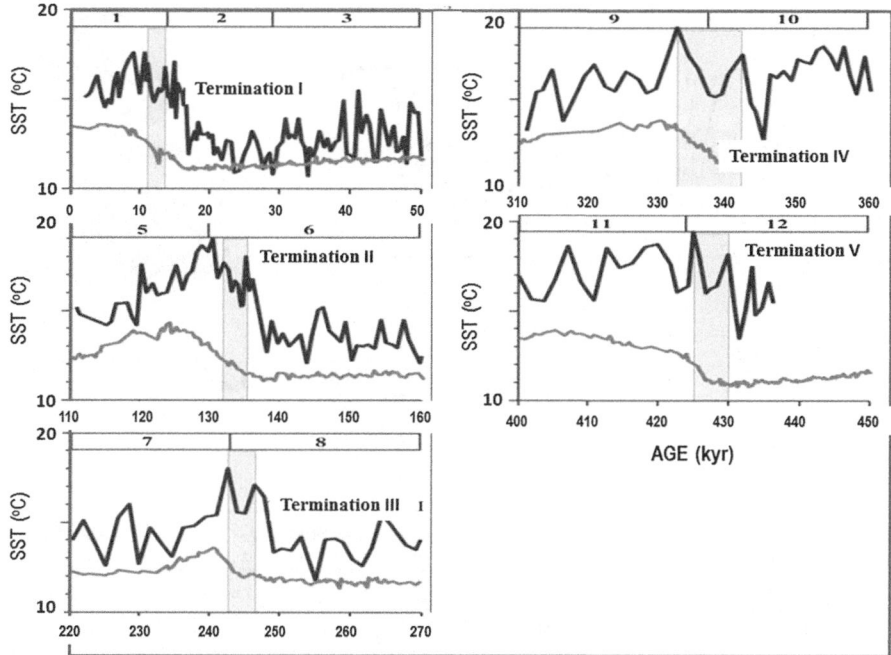

Fig. 3.8 Evolution of sea surface temperatures in 5 glacial-interglacial transitions recorded in ODP 1089 at the sub-Antarctic Atlantic Ocean. Grey lines—$\delta^{18}O$ measured on Cibicidoides plankton; Black lines—sea surface temperature. Marine isotope stage numbers are indicated on top of diagrams. Note the stadials following interglacial peak temperatures, analogous to the Younger dryas (12.9–11.7 kyr) preceding the onset of the Holocene (Cortese et al. 2007, Fig. 4; John Wiley and Sons, by permission)

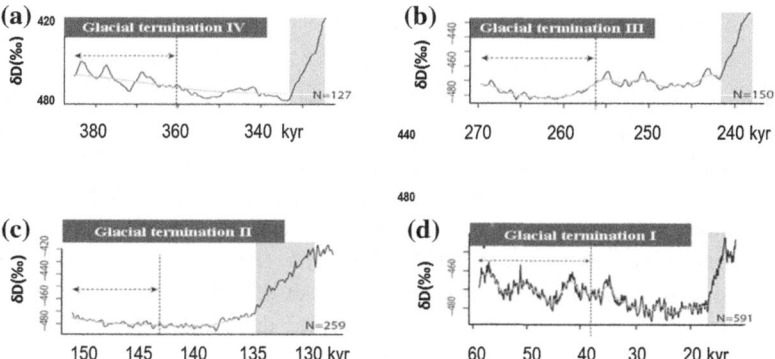

Fig. 3.9 Reconstructions of abrupt climate shifts during glacial-interglacial cycles: **a** glacial termination IV; **b** Glacial termination III; **c** glacial termination II; **d** glacial termination I. Note the slowdown in the dynamics of the system prior to the transition. Data from the GISP-2 ice core, ODP Hole 658C and Antarctica Vostok ice core (Dakos et al. 2008, Fig. 1; Proceeding of the National Academy of Science, by permission)

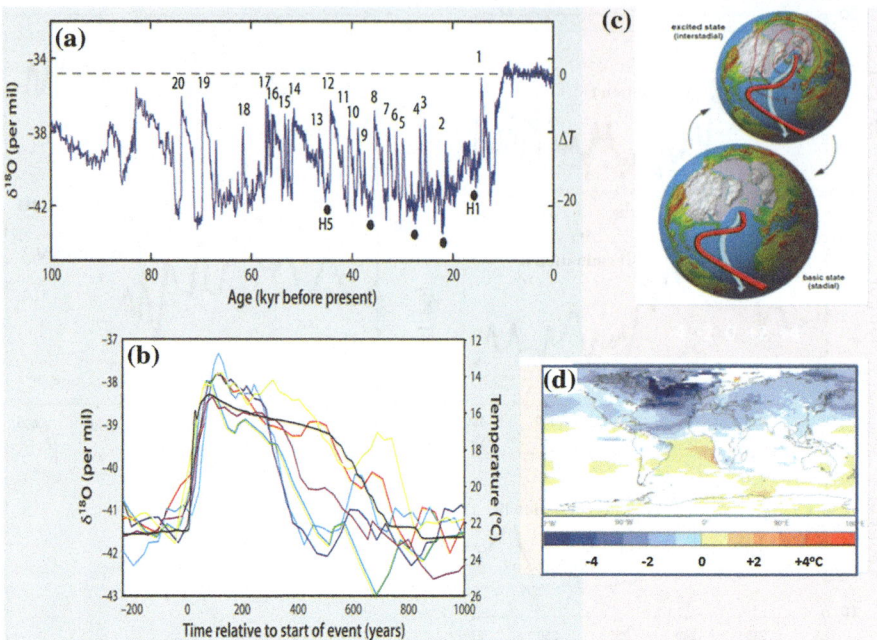

Fig. 3.10 Intra-glacial Dansgaard-Oeschger cycles. **a** Temperature variations during the last glacial period since 100 Kyr, displaying 21 Dansgaard-Oeschger interstadial cycles recorded by $\delta^{18}O$ values from the GRIP ice core of the Greenland Ice Sheet; **b** A single Dansgaard-Oeschger cycle. The black line shows a model simulated D/O event (Rahmstorf and Stocker, 2004; Springer, by permission); **c** phases of the North Atlantic thermal current during the warm excited state (interstadial, upper globe) and the cold basic state (stadial, lower globe) of the D-O cycle (Ganopolski and Rahmstorf 2002, Figs. 1 and 2; Springer, by permission); **d** Changes in surface air temperature caused by a stadial state shutdown of North Atlantic Deep Water (NADW) formation in a current ocean–atmosphere circulation model (Rahmstorf 2002, Fig. 1; Nature, by permission). Note the extreme rate of temperature rise during the D-O cycles

Fastest rates of temperature rise are recorded over ~1,500 years-long cycles (Dansgaard-Oeschger [D-O] cycles) (Fig. 3.10) during the last glacial period of ~75 to 20 kyr, superposed on longer-term cooling trends and culminating in Heinrich events (Yokoyama and Esat 2011). The cycles were related to melting of the Greenland and Laurentian ice sheets, with effects extending to the North Atlantic Thermohaline Current (NATH) (Fig. 3.10) (Broecker 2000). Temperature rises recorded in Greenland ice cores at the outset of D-O cycles reach ~6–8 °C (Ganopolski and Rahmstorf 2002) within several decades, suggestive of ~3 to 4 °C mean global changes. Approximate mean global temperature rise rates are estimated in the range of ~0.01 to 0.2 °C/year, commensurate with, or higher than, 20–21st centuries temperature rise rates (Table A2). CO_2 increases associated with D-O cycles are estimated as ~ 20 ppm, with CO_2 rise rates of ~0.2 ppm/year, an order of magnitude less than modern rates (Table A2).

3 Cenozoic Atmospheres and Early Hominins

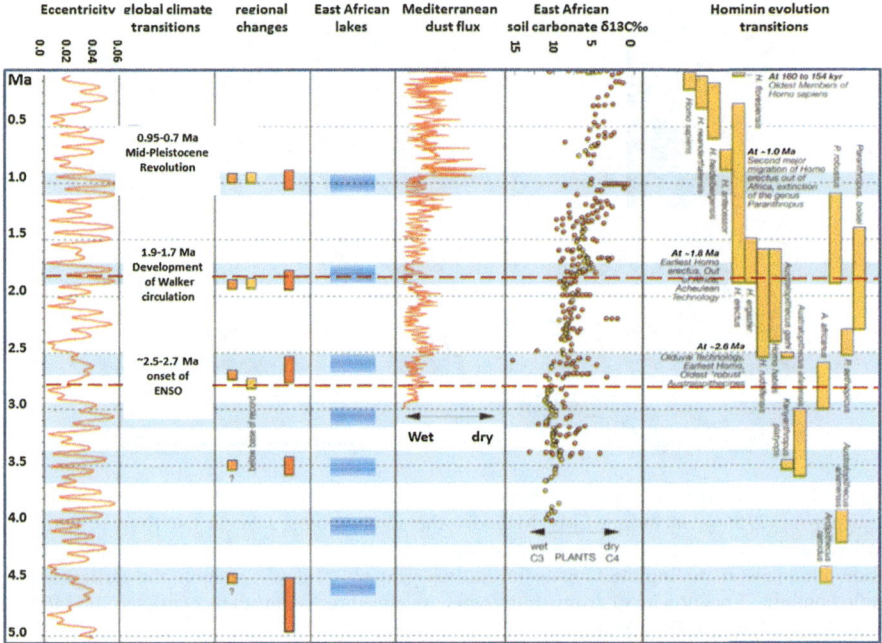

Fig. 3.11 A tabulated summary of climate and hominin evolution from 5.0 Ma-ago, including comparisons of eccentricity variations, high-latitude climate transitions, Mediterranean dust flux, soil carbonate carbon isotopes, East African lake occurrences, Hominins species appearances and durations (Maslin and Trauth 2006, Fig. 13.2; Springer, by permission)

Abrupt climate shifts including stadial cooling during the interglacials (Fig. 3.8) are exemplified by (1) the onset and termination of the 12.9–11.7 kyr *Younger Dryas* cold phase over periods as short as 1–3 years (Steffensen et al. 2008), and (2) a sharp temperature decline by several degrees Celsius at ∼8.2 kyr in the North Atlantic, associated with discharge of cold water from the Laurentian ice sheet through Lake Agassiz (Wagner et al. 2002; Lewis et al. 2012) (Fig. 7.3).

The transitions through the Pliocene and Pleistocene from tropical to savannah environments in Africa, accompanied with faunal diversification from tropical species toward arid-zone type species, including Antelopes (Bovids) about 2.8, 1.8–1.7 and 0.8–0.7 Ma (Fig. 6.1), signify an increase in climate variability and enhanced pace of evolution. Glacial termination events, preceded by low-variability lulls (Dakos et al. 2008) (Fig. 3.9), involved high rates of greenhouse gas rise and of temperature rise on the order of 0.0004 °C/year and 0.009 ppm CO_2/year, faster than Miocene and Pliocene rates but lower by orders of magnitude compared to modern rates (Table A2).

Human evolution associated with these climate transitions was expressed by diversification and appearance of Olduvan stone tools from about ∼2.7 Ma and

Fig. 3.12 East African rift valleys (NASA; http://earthobservatory.nasa.gov/IOTD/view.php?id=77566), including Lake Magadi—the southernmost lake in the Kenyan Rift Valley in a catchment of faulted volcanic rocks. The area is rich in fossils of the hominin ancestors of modern humans, including the famous 3.2-million-year-old "Lucy" skeleton, the earliest known adult hominin. Location map from: http://www.sciencedirect.com/science/article/pii/S1464343X05001251 (Elsevier, by permission)

Acheulean stone tools from about ~ 1.7 Ma (deMenocal 2004; Klein 2009) (Figs. 3.11, 6.2). Survival stresses associated with extreme glacial-interglacial climates from about 0.9 Ma saw near-tripling of the Homo cranial cavity from ~ 450 to $\sim 1{,}200$–$1{,}500$ cc, mastering of fire about or later than 2 Ma and, from about ~ 160 kyr, cultural developments including burial, ornamentation and rock painting by Homo sapiens.

References

Alvarez LW, Alvarez W, Asaro F, Michel HV (1980) Extra-terrestrial cause for the cretaceous-tertiary extinction: experimental results and theoretical interpretation. Science 208:1095–11086
Bard E, Frank M (2006) Climate change and solar variability: what's new under the sun? Earth Planet Sci Lett 248:1–14
Beerling DJ, Royer D (2011) Convergent cenozoic CO2 history. Nat Geosci 4:418–420
Beerling DJ, Lomax BH, Royer DL, Upchurch GR, Kump LR (2002) An atmospheric pCO_2 reconstruction across the cretaceous-tertiary boundary from leaf mega fossils. Proc Nat Acad Sci 99:7836–7840
Berger WH, Jansen E (1994) Mid-pleistocene climate shift: the Nansen connection. In: Johannessen O, Muench R, Overland J (eds) The polar oceans and their role in shaping the global environment, vol 85., Geophys MonoAmerican Geophysical Union, Washington, DC, p 295–311

References

Broecker WS (2000) Abrupt climate change: causal constraints provided by the paleoclimate record. Earth Sci Rev 51:137–154

Browning JV, Miller KG, Pak DK (1996) Global implications of lower to middle eocene sequence boundaries on the New Jersey coastal plain: the icehouse cometh. Geology 24:639–642

Chandler M, Dowsett H, Haywood A (2008) The PRISM model/data cooperative: mid-pliocene data-model comparisons. PAGES News 16(2):24–25

Cortese G, Abelmann A, Gersonde A (2007) The last five glacial-interglacial transitions: a high-resolution 450,000-year record from the subantarctic Atlantic. Paleoocean 22:PA4203

Cui Y, Kump LR, Ridgwell AJ, Charles AJ, Junium CK, Diefendorf AF, Freeman KH, Urban NM, Harding IC (2011) Slow release of fossil carbon during the palaeocene–eocene thermal maximum. Nature Geosci 4:481–485

Dakos V, Scheffer M, Van Nes EH, Brovkin V, Petoukhov V, Held H (2008) Slowing down as an early warning signal for abrupt climate change. Proc nat Acad Sci 105:14308–14312

Deino AL, Kingston JD, Glen JM, Edgar RK, Hill A (2006) Precessional forcing of lacustrine sedimentation in the late Cenozoic Chemeron basin, central Kenya rift, and calibration of the Gauss/Matuyama boundary. Earth Planet Sci Lett 247:41–60

deMenocal PB (2004) African climate change and faunal evolution during the Pliocene-Pleistocene. Earth Planet Sci Lett 220:3–24

EPICA Community Members (2004) Eight glacial cycles from an Antarctic ice core. Nature 429:623–628

Eyles N (1993) Earth's glacial record and its tectonic setting. Earth Sci Rev 35:1–248

Feakins SJ, deMenocal PB, Eglinton TI (2005) Biomarker records of late neogene changes in Northeast African vegetation. Geology 33:977–980

Fedorov AV, Dekens PS, McCarthy M, Ravelo AC, deMenocal PB, Barreuri M, Pacanowski RC, Philander SG (2006) The pliocene paradox. Science 312:1485–1489

Frakes LA, Francis JE, Syktus JI (1992) Climate modes of the phanerozoic. Cambridge University Press, Cambridge

Ganopolski A, Rahmstorf S (2002) Abrupt glacial climate changes due to stochastic resonance. Physics Rev Lett 88:3–6

Glikson AY, Jablonski D, Westlake S (2010) Origin of the Mt Ashmore structural dome, west Bonaparte basin, Timor Sea. Aust J Earth Sci 57:411–430

Hansen J, Sato M, Kharecha P, Lea DW, Siddall M (2007) Climate change and trace gases. Phil Trans Roy Soc 365A:1925–1954

Hansen J, Sato M, Kharecha P, Beerling D, Masson-Delmotte V, Pagani M, Raymo M, Royer DL, Zachos JC (2008) Target atmospheric CO_2: where should humanity aim? Open Atmos Sci J 2:217–231

Klein R (2009) The human career: human biological and cultural origins. University of Chicago Press, Chicago

Lewis CFM, Miller AAL, Levac E, Piper DJW, Sonnichsen GV (2012) Lake Agassiz outburst age and routing by labrador current and the 8.2 ka cold event. Quatern Int 260:83–97

Maslin MA, Seidov D, Lowe J (2001) Synthesis of the nature and causes of sudden climate transitions during the Quaternary. In: Seidov D, Haupt BJ, Maslin M (eds) The Oceans and Rapid Climate Change: Past, Present and Future. Am. Geophys. Union Geophys. Monogr. Series 126:9–52

Maslin MA, Trauth MH (2006) Plio-Pleistocene east african pulsed climate variability and its influence on early human evolution. In: Grine GE, Fleagle JG, Leakey RE (eds) Contributions from the third stony brook human evolution symposium and workshop 3–7 Oct

Maslin MA, Christensen B (2007) Tectonics, orbital forcing, global climate change, and human evolution in Africa: introduction to the African paleoclimate special volume. J Human Evol 53(5):443–464

Maslin MA, Trauth MH (2009) Plio-Pleistocene East African pulsed climate variability and its influence on early human evolution. In: The first humans: origin and early evolution of the genus homo. Verteb paleobiology paleoanthropology, p 151–158

Miller KG, Wright JD, Katz ME, Wade BS, Browning JV, Cramer BS, Rosenthal Y (2009) Climate threshold at the Eocene-Oligocene transition: Antarctic ice sheet influence on ocean circulation. In: Koeberl C, Montanari A (eds) The late eocene earth—hothouse, icehouse, and impacts: geological society of American Sp. papers, vol 452, p 1–10

Overpeck J, Bette T, Otto-Bliesner L, Gifford H, Mille M, Daniel RM, Alley RB, Kiehl JT (2006) Paleoclimatic evidence for future ice-sheet instability and rapid sea-level rise. Science 311:1747–1750

Pearson PN, Foster GL, Wade BS (2009) Atmospheric carbon dioxide through the eocene–oligocene climate transition. Nature 461:1110–1113

Petit JR et al (1999) 420,000 years of climate and atmospheric history revealed by the Vostok deep Antarctic ice core. Nature 399:429–436

Pollard D, DeConto RM (2005) Hysteresis in cenozoic Antarctic ice sheet variations. Glob Planet Change 45:9–21

Potts R (1998) Environmental hypothesis of hominin evolution. Yearbook Phys Anthrop 41:93–136

Rahmstorf S (2002) Ocean circulation and climate over the last 120,000 years. Nature 419:6903

Rahmstorf S, Stocker TF (2004) Living with global change: consequences of changes in the earth system for human well-being.In: Steffen W (ed) Box 5.6 in: a planet under pressure—global change and the earth system. Springer, Berlin, p 240–241

Roe G (2006) In defence of Milankovitch. Geophys Res Lett 33:L24703

Royer DL (2006) CO_2-forced climate thresholds during the phanerozoic. Geochim Cosmochim Acta 70:5665–5675

Royer DL, Berner RA, Montañez I, Neil P, Tabor J, Beerling DJ (2004) CO_2 as a primary driver of phanerozoic climate. GSA Today 14:3

Ruddiman WF (1997) Tectonic uplift and climate change. Plenum Press, New York, p 535

Ruddiman WF (2008) Earth's climate, past and future, 2nd edn. WH Freeman, New York. ISBN 978-0-7167-8490-6

Solanki SK (2002) Solar variability and climate change: is there a link? Sol Phys 43:59–513

Steffensen JP et al (2008) High-resolution greenland ice core data show abrupt climate change happens in few years. Science 321:680–684

Teaford MF, Ungar PS (2000) Diet and the evolution of the earliest human ancestors. Proc Nat Acad Sci USA 97:13506–13511

Trauth MH, Maslin MA, Deino AL, Strecker MR, Bergner AGN, Duhnforth M (2007) High- and low-latitude forcing of Plio-Pleistocene East African climate and human evolution. J Hum Evol 53:475–486

Trauth MH, Maslin MA, Deino AL, Junginger A, Lesoloyia M, Odada EO, Olago DO, Olaka LA, Strecker MR, Tiedemann R (2010) Human evolution in a variable environment: the amplifier lakes of Eastern Africa. Quater Sci Rev 29:2981–2988

Wagner F, Aaby B, Visscher H (2002) Rapid atmospheric CO2 changes associated with the 8,200-years-B.P. cooling event. Proc Nat Acad Sci 99:12011–12014

Yokoyama Y, Esat TM (2011) Global climate and sea level: enduring variability and rapid fluctuations over the past 150,000 years. Oceanography 24:54–69

Zachos JC, Breza JR, Wise SW (1992) Early oligocene ice-sheet expansion on Antarctica–stable isotope and sedimentological evidence from Kerguelen Plateau, Southern Indian Ocean. Geology 20:569–573

Zachos J, Pagani M, Sloan L, Thomas E, Billups K (2001) Trends, rhythms, and aberrations in global climate 65 Ma to present. Science 292:686–693

Zachos J, Dickens GR, Zeebe RE (2008) An early cenozoic perspective on greenhouse warming and carbon-cycle dynamics. Nature 451:279–283

Part II
The Great Mass Extinctions of Species

If the radiance of a thousand suns were to burst at once into the sky, that would be like the splendor of the mighty one

Bhagavad-Gita. Robert Oppenheimer at the Trinity atomic test

Chapter 4
Mass Extinction of Species

Abstract Early conflicts between uniformitarian and gradual theories of evolution (James Hutton: 1726–1797; Charles Lyell: 1797–1875) and catastrophic theory (Cuvier: 1769–1832) have been progressively resolved by advanced paleontological, sedimentary, volcanic and asteroid impact studies and by paleo-climate studies coupled with precise isotopic age determinations, indicating periods of gradual evolution were interrupted by abrupt events which have transformed the habitat of plants and organisms and resulted in mass extinction of species.

The geological record betrays a close correspondence between paleontological, sedimentary, volcanic, asteroid impact and paleo-CO_2 and paleo-temperature trends, allowing identification of environmental factors which underlie the evolution and extinction of species (McElwain et al. 1999, 2007; Beerling 2002a, b; Beerling et al. 2002; Keller 2005; Glikson 2005). Five major mass extinction events and several moderate extinction events have affected the evolution of marine invertebrate (Fig. 4.1). High-resolution regional palaeoecological studies indicate extensive ecological upheaval, high species-level turnover and recovery intervals lasting millions of years, with close correlations to upheavals affecting terrestrial vegetation (McElwain and Punyasena 2007).

When a large (>200 m) asteroid hits a solid surface at a high angle it penetrates to a depth of approximately x1.5 times its diameter, depending on the rheology of the impacted rocks, where its kinetic energy is translated into heat, triggering an explosion, fragmentation, cratering, melting and vaporization of the immediately surrounding rocks. In craters larger than about 4 km the Earth's crust rebounds to form a central uplift (French 1998; Glikson et al. 2013a, b). Depending on the size of the impact, seismic waves propagate, leading to earthquakes, faulting and tsunami waves over large regions. Environmental effects of asteroid impacts include the initial fireball flash, mega-tsunami waves, release of aerosols (dust, sulfur dioxide, carbon soot), acid rain and release of greenhouse gases (water, CO_2, methane, nitrous oxide) from cratered regions, leading to ocean acidification. This leads to an asteroid winter phenomenon, with some 10–20 % of solar radiation blocked for 8–13 years (Pope et al. 1997), followed by a greenhouse-gas induced warm period lasting centuries to millennia. Species which have escaped

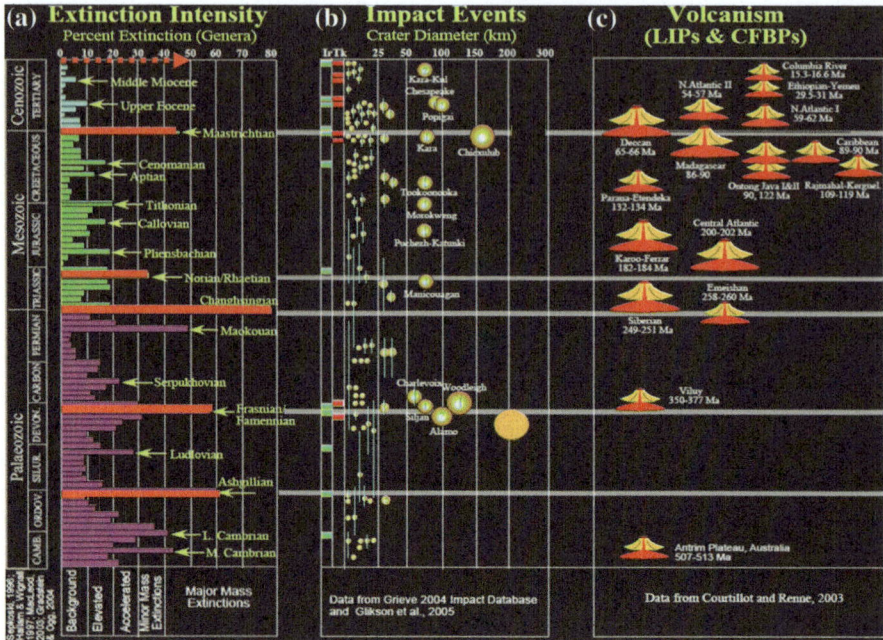

Fig. 4.1 Phanerozoic mass extinctions, asteroid impacts, and large igneous provinces: **a** Extinction intensity; **b** Impact events; **c** Volcanism. Stratigraphic subdivisions and numerical ages are after Gradstein and Ogg (2004). The extinction record is based on genus-level data by Sepkoski (1996). The number of impact events, size and age of craters follows largely the Earth Impact Database (2005 http://www.unb.ca/fredericton/science/research/passc/), with modification by the author (Glikson and Uysal 2013b) (*Courtesy* G. Keller)

the immediate regional and transient effects of large impacts were affected by the long-lasting consequences of the well-mixed greenhouse gases, mainly CO_2, Nitrous oxide and methane. CO_2 stays in the atmosphere for thousands to tens of thousands of years, leading to extended periods of high global temperatures, compounding the effects on the biosphere. Effects on the oceans include acidification, anoxia (oxygen solubility decreasing with higher temperatures) and consequent anoxic conditions and toxic H_2S emanations (Ward 2007).

Similar although more protracted effects can be induced by large scale longer-term volcanic eruptions. The following temporal associations between volcanic and/or asteroid events and mass extinction are relevant (Fig. 4.1; Tables A1 and A2):

1. Acraman/Bunyeroo ∼580 Ma asteroid impact and Acritarch radiation (Grey et al. 2003; Grey 2005).
2. Late Devonian (Frasnian-Fammenian) ∼374 Ma, possible asteroid impacts and mass extinction.

3. End-Devonian ~360 Ma impact cluster (Woodleigh, Siljan, Charlevoix, Alamo, possibly Warburton East) and the destruction of reefs (McGhee 1996; Balter et al. 2008).
4. Permian–Triassic boundary ~251 Ma volcanic (Norilsk) and asteroid impact (Araguinha) events and mass extinction of species (Renne et al. 1995; Ross and Ross 1995; Wignall and Twitchett 1996; Twitchett et al. 2001; Racki 2003; Ward 2007).
5. Late-Triassic ~216 Ma impact (Manicouagan, Rochechouart) and mass extinction.
6. End-Triassic ~200 Ma opening of the Atlantic Ocean, extensive volcanism and extinction (Olsen and Sues 1986; McElwain et al. 1999; Jourdan et al. 2009; Whiteside et al. 2010).
7. Early Jurassic (Pliensbachian) ~183 Ma Karoo volcanism and extinction.
8. End-Jurassic ~145 Ma impact cluster (Morokweng, Gosses Bluff, Mjolnir), opening of the Indian Ocean and extinction (McElwain et al. 1999).
9. Cretaceous-Tertiary boundary ~65 Ma impacts (Chicxulub, Boltish), Deccan volcanism and mass extinction.

However, precise age correlations between impacts and extinctions have only been established to date regarding events [1] and [9]. Intermediate scale Extinctions also occur at the Middle Cambrian, Late Cambrian and end-Silurian (Fig. 4.1). The Cambrian extinctions are likely related to the upper Cambrian Kalkarindji continental flood basalt volcanism (Glass and Phillips 2006).

4.1 Acraman Impact and Acritarchs Radiation

The ~580 Ma Acraman impact structure, estimated as ~90 km in diameter (Gostin et al. 1986; Gostin and Zbik 1999; Williams et al. 1996; Williams and Gostin 2005), and its ejecta layer found up to 550 km away from the crater, postdate the Marinoan glaciation (650–635 Ma). The impact was closely followed by radiation of Acritarch phytoplanktons, including an abrupt change from Ediacaran leiosphere palynoflora (ELP) to Ediacaran complex Acritarchs palynoflora (ECAP), presenting the oldest example of biological radiation following large catastrophic events (Grey et al. 2003; Grey 2005) (Fig. 4.2). The sequence from the terminal glacial sediments of the Cryogenian (~635 Ma) to the base Cambrian includes (1) cap carbonates, representing likely greenhouse gas-driven glacial collapse; (2) clastic sediments; (3) the ~580 Ma Acraman impact ejecta overlain by the Acritarchs-radiation horizon; (4) Ediacara fauna ~550 Ma (Fig. 1.10) and (5) ~544 Ma base Cambrian. The Acraman event is associated with marked negative $\delta^{13}C$ anomalies which signify increased deposition of organic matter (Calver 2000; Walter et al. 2000).

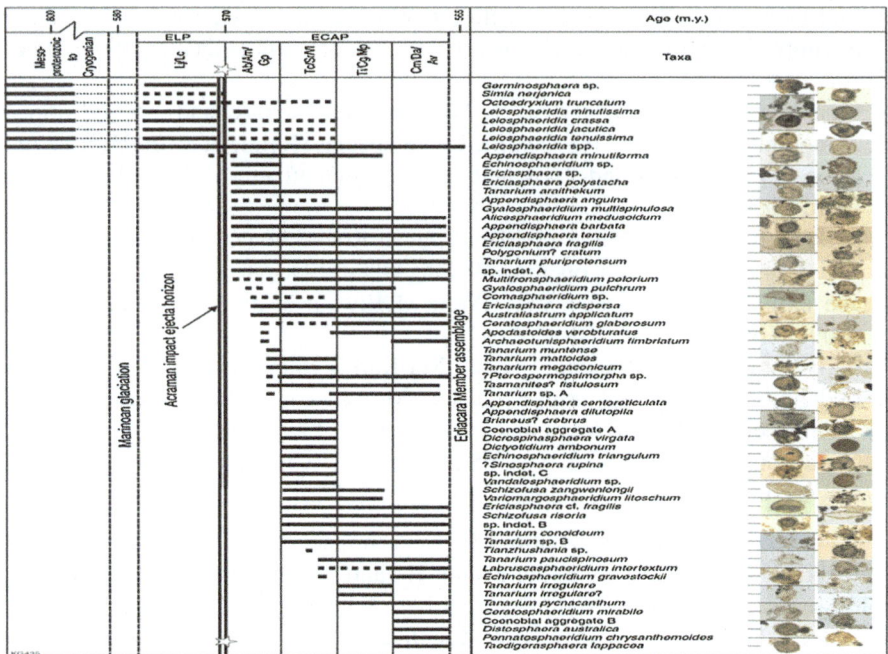

Fig. 4.2 Stratigraphic distribution of Acritarchs following the Marinoan glaciation, displaying a major discontinuity at ~570 Ma coinciding with ejecta from the Acraman impact event, followed by major radiation; *ELP* Ediacaran Leiosphere Palynoflora; *ECAP* Ediacaran Complex Acanthomorph Palynoflora. (Grey 2005); (*Courtesy* K. Grey and of the Geological Survey of Western Australia, Department of Mines and Petroleum, Western Australia 2013

4.2 Late Ordovician Mass Extinction

The end-Ordovician period, marked by a glaciation about ~445.6–443.7 Ma and possibly longer (Frakes et al. 1992), saw two phases of extinction involving ~57 % of genera (Hallam and Wignall 1997), including pelagic graptolites and most benthic groups (trilobites, brachiopods, bryozoans, echinoderms). Factors involved in the first phase of extinction included cooling, glaciation, sea-level regression and major changes in oceanic circulation, leading to extinction of pelagic groups including graptolites and conodonts. The second phase appears to have been related to warming and ocean bottom anoxia eliminating shelf habitats (Hallam and Wignall 1997; Keller 2005). According to Kump et al. (1999) CO_2 levels declined during the glaciation from 5,000 to 3,000 ppm, high levels compensated by low insolation about 4 percent lower than at present level of 342 Watt/m^2.

4.3 Late and End-Devonian Mass Extinctions

Possible factors associated with late Devonian mass extinctions include volcanism of the Viluy Traps, East European platform, estimated as >510,000 km^3 and dated in the range 377 and 350 Ma (Keller 2005) (Fig. 4.3). The end-Devonian at ∼360 Ma was marked by a large asteroid impact cluster including Woodleigh (D = 120 km), Alamo (D = 100 km), Charlevoix (D = 54 km) and Siljan (D = 52 km) and possibly Warburton (D > 400 km) (Glikson et al. 2013a, b). Devonian mass extinction events (McGhee 1996; Hallam and Wignall 1997) include a ∼387 Ma extinction (∼30 % of Genera) and ∼374 Ma extinction (58 % of Genera) affecting pelagic fauna (Ammonoids, Cricoconaids, Placoderms, Conodonts, Agnathans) and benthic groups (Rugose corals, Trilobites, Ostracods and Brachiopods). The extinction involved collapse of Stromatoporoid reefs (Keller 2005). End-Devonian ∼359 Ma extinction (∼30 % of Genera) affected fish (Placoderms), ammonoids, conodonts, stromatoporoids, rugose corals, trilobites and ostracods. Major factors included ocean anoxia, declining biological activity (high δ^{13}C), and warming (low δ^{18}O) (Balter et al. 2008). The late Devonian mass extinctions are superposed on a protracted cooling trend associated with a decline in CO_2 levels from a range of ∼3,200–5,200 ppm to below ∼500 ppm. Concomitant decline in δ^{13}C from ∼22 to ∼15 ‰ from ∼405 Ma to 280 Ma is indicated by paleosols (Mora et al. 1996). The development in the Late Devonian of plant megaphyll leaves with their branched veins containing high stomata density allowed vegetation to adapt to the cool low-CO_2 conditions of the Carboniferous-Permian (Rothwell et al. 1989; Beerling et al. 2001).

4.4 Late Permian and Permian–Triassic Mass Extinctions

Major eruptions of the Siberian Norlisk magmatic province and Emeishan volcanism (Renne et al. 1995; Wignall and Twitchett 1996; Wignall 2001) about ∼251 Ma (251.7 ± 0.4 to 251.3 ± 0.3 m.y., Kamo et al. 2003) and a large asteroid impact (Araguinha, Brazil, D = 40 km, ∼247.8 ± 3.8 Ma; Tohver et al. 2012), which excavated carbonates and shale (Table A2), has led to a rise of atmospheric CO_2 levels to ∼3,400 ppm (Royer 2006) (Fig. 4.4a), associated with the greatest mass extinction recorded in geological history (Fig. 4.4b).

Two major extinction phases are defined:

(a) ∼50 % of genera extinguished at the ∼268–265 Ma Late Permian Maokouan Stage. Tropical zones saw the extinction of echinoderms, corals, brachiopods, sponges, fusulinid foraminifers and ammonoids (Ross and Ross 1995; Keller 2005) (Table A1; Fig. 4.4b).

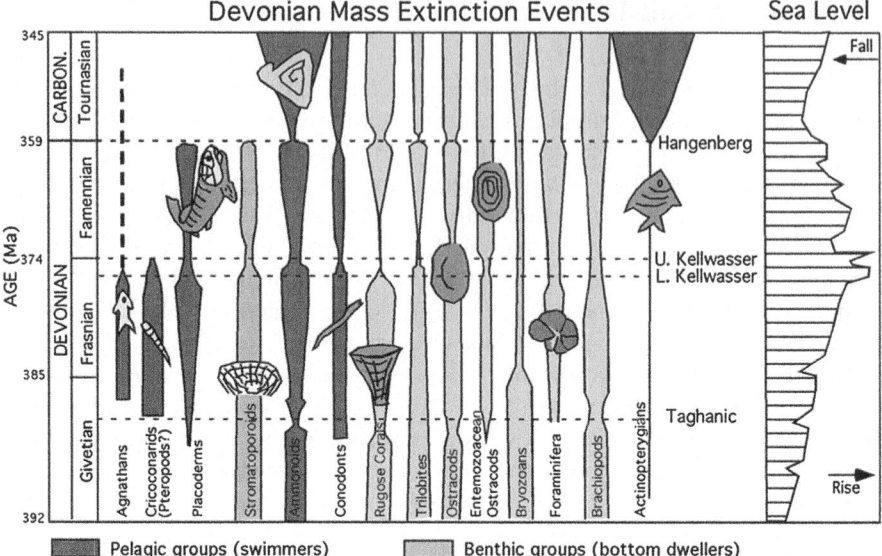

Fig. 4.3 Biotic effects of Devonian mass extinction events (Hallam and Wignall 1997). Major mass extinctions are in the *lower* and *upper* Kellwasser at the end of the Frasnian, decimating reef system, shallow benthic fauna and pelagic swimmers. The crises were associated with sea-level rises, warm climate and widespread ocean anoxia. The Hangenberg extinction affected mainly pelagic groups including ammonoid and fish. The mid-Devonian Taghanic extinction was part of a long-term diversity decline (after Keller 2005, Fig. 21; Australian Journal of Earth Science, by permission)

(b) ~78 % of genera extinguished at the ~251 Ma end-Permian Changhsingian Stage, affecting abrupt extinction of the Rugose Corals, Bryozoans, complex Foraminifera, many Gastropod and Bivalve families, radiolarians and many Ammonoid families (Hallam and Wignall 1997; Racki 2003) (Table A1, Fig. 4.4b).

The two events were separated by the Capitanian and Wuchiapingian Stages (265.8–253.8 Ma) (Keller 2005). An abrupt nature of these events is indicated by their short duration of 10–50.10^3 years and negative $\delta^{13}C$ excursion indicating deposition of fauna and flora remains (Twitchett et al. 2001). Nektonic (free swimming) fauna, including fish, conodonts and nautiloids survived better thanks to their mobility in the upper water column above anoxic bottom water (Keller 2005). Anoxia is evidenced by sulphide-rich and black clay sediments and negative $\delta^{13}C$ anomalies testifying to mass settling of organic matter. Grasby et al. 2011, suggested a link between extinction and a release of carbon ash/char derived from the combustion of Siberian coal and organic-rich sediments by flood basalts, which was dispersed globally and created toxic marine conditions.

Berner (2005) investigated geochemical trends across the Permian–Triassic boundary from isotopic $\delta^{13}C$ and $\delta^{34}S$ mass balance and estimates of weathering and

4.4 Late Permian and Permian–Triassic Mass Extinctions

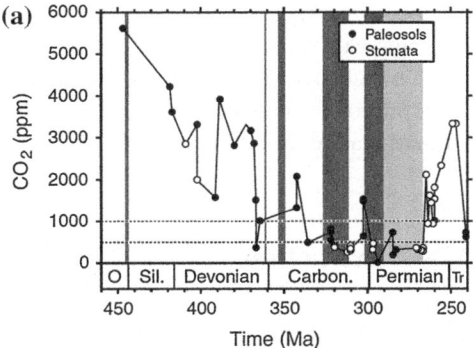

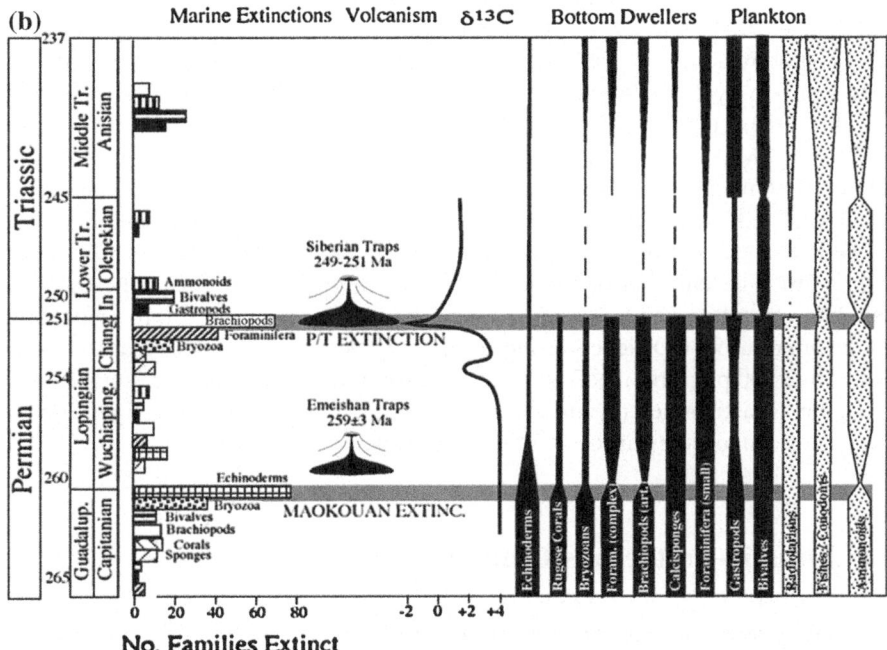

Fig. 4.4 a CO_2 and temperature records for the Late Ordovician to early Triassic (460–240 Ma). Note the upper to late Devonian and Permian–Triassic boundary peaks of ∼3,000 ppm CO_2, corresponding to a major mass extinction of species (Royer 2006, Fig. 3A; Elsevier, by permission) **b** Faunal turnover, impacts and volcanism across the Permian–Triassic transition. Faunal data modified after Hallam and Wignall (1997); volcanism after Courtillot and Renne (2003). (Keller 2005, Fig. 20; Australian Journal of Earth Science, by permission)

burial of carbon and sulfur. A drop in the rate of organic burial from the late Permian to the mid-Triassic is attributed to rising aridity and decrease in biomass due to a transition from forests to herbaceous grassland. A major drop in oxygen from 30 to 13 percent was associated with an increase in the ratio of pyrite to organic carbon and in development of marine euxinic basins. Consequences included extinction of vertebrates and loss of giant insects and amphibians. According to Ward (2007) ocean acidification due to rising CO_2 levels, polar ice melt, weakened ocean current system and the conveyor belt which provides oxygen, consequent anoxia, production of H_2S by sulphur-reducing microbes and its release to the atmosphere, constituted critical factors in sea and land mass extinction.

4.5 End-Triassic Mass Extinction

The opening of the Central Atlantic magmatic province by the end-Triassic at ~ 200 Ma, involving copious basaltic volcanism (Hames et al. 2003; Courtillot and Renne 2003; Jourdan et al. 2009), affected a major mass extinction event represented by a large negative carbon isotope excursion, reflecting perturbations of the carbon cycle including an increase in CO_2 (Beerling 2002a, b; Whiteside et al. 2010) (Fig. 4.5a). The end-Triassic was preceded by a Norian ($\sim 216–213$ Ma) extinction associated with the large Manicouagan impact (D ~ 100 km; 214 ± 1 Ma) (Table A1). The extinction affected ammonites, which radiated back in the early Jurassic, reef organisms, conodonts and bivalves, as well as a crisis in terrestrial plants (Hallam and Wignall 1997; Keller 2005) (Fig. 4.5b). The duration of the extinction is variously estimated as between 50 and 200 kyr (Olsen and Sues 1986). According to Beerling (2002a, b), depending on the proxy used, CO_2 levels rising from the Rhaetian (~ 204 Ma) reached about $\sim 1,300–2,200$ ppm from leaf stomata and a wider range from carbon isotopes, just above the Rhaetian-Hettangian (early Jurassic) boundary (Fig. 4.5a), signifying an extreme greenhouse event of ~ 34 % of genera.

4.6 Jurassic-Cretaceous Climate Anomalies

CO_2 and temperature records for the Jurassic to late Cretaceous ($\sim 200–80$ Ma) indicate strong climate fluctuations, expressed by variations in atmospheric CO_2 levels in the order of 2000–5000 ppm (Royer 2006) (Fig. 4.6). This resulted in sharp changes between cool events and warm periods, the latter likely related to the opening of oceanic gaps, breakdown of Gondwana and associated intense volcanism, including Karoo volcanism (~ 183 Ma) during this era.

4.7 K–T (Cretaceous-Tertiary Boundary) Mass Extinction

The K–T boundary (64.98 ± 0.05 Ma) marks the 2nd largest mass extinction of species recorded in Earth history, causing the disappearance of 46 percent of living genera (Keller 2005) (Figs 4.7a, b). Alvarez et al. (1980) discovered a hiatus

4.7 K–T (Cretaceous-Tertiary Boundary) Mass Extinction

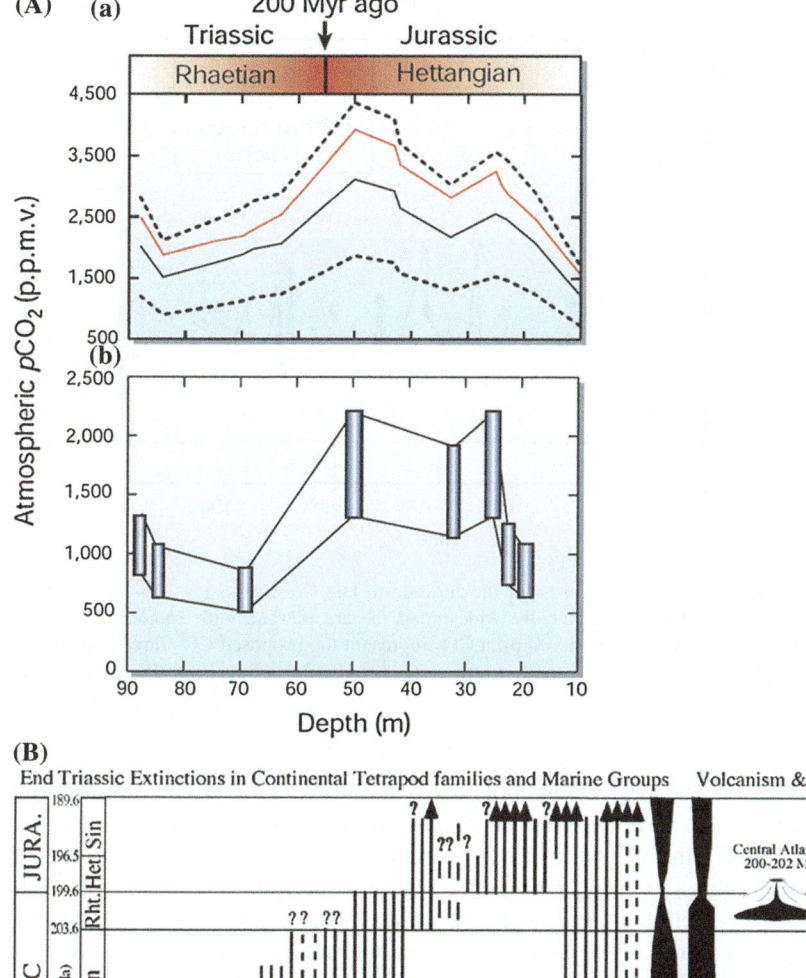

Fig. 4.5 A Palaeo-atmospheric CO_2 variations across the Triassic–Jurassic boundary a Atmospheric CO_2 changes calculated using $\delta^{13}C$ values and a constant Triassic paleosol carbonate value up to the Triassic–Jurassic boundary and a constant Jurassic paleosol value; b Atmospheric CO_2 changes reconstructed from the stomata of fossil leaves. *Vertical bars* denote the *upper* and *lower range* for any given depth calculated using this technique (Beerling 2002a Fig. 1; Nature, by permission) B Mass extinction, impacts and volcanism across the Triassic–Jurassic transition. Fauna changes modified after Hallam and Wignall (1997); volcanism after Courtillot and Renne (2003). (Keller 2005, Fig. 19; Australian Journal of Earth Science, by permission)

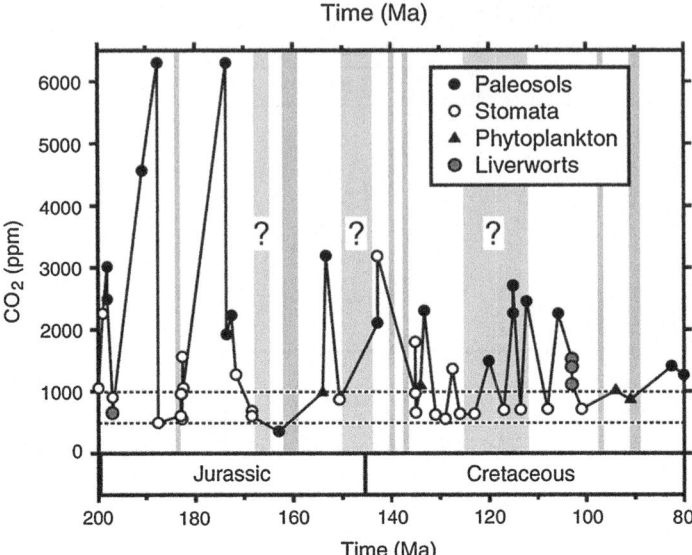

Fig. 4.6 CO_2 and temperature records for the Jurassic to late Cretaceous ($\sim$200–80 Ma). Cold periods with evidence for geographically widespread ice are marked with *shaded bands*. The *horizontal dashed lines* at 1,000 and 500 ppm CO_2 represent the proposed CO_2 thresholds for the initiation of globally cool events and warm periods (Royer 2006, Fig. 3B; Elsevier, by permission)

across the Cretaceous-Paleocene boundary in Italy, where a foraminifera-rich white limestone facies containing large-scale Globotruncana contusa is abruptly replaced by overlying clay-rich red limestone termed Scaglia Rossa containing smaller foraminifera (Globigerina eugubina) and micron-scale algal coccoliths (Fig. 4.7a). At the classic locality at Gubbio a $\sim$1 cm-thick boundary clay layer consisting of a lower $\sim$5 mm-thick grey clay zone and an upper $\sim$5 mm-thick red clay zone termed 'fire layer'—contains an Iridium anomaly of up to $\sim$9 ppb. Similar relations are observed in numerous localities around the world (Fig. 4.7b). The boundary coincides with a major geomagnetic reversal correlated with a marine magnetic anomaly sequence dated with foraminifera. The parent craters of the K–T event have been identified, including Chicxulub (170 km-diameter, Yucatan Peninsula, Mexico) and Boltysh ($\sim$25 km-diameter; 65.17 $\pm$ 0.64 Ma, Ukraine). Since the initial discovery of K–T impact ejecta, best preserved in deep water environment, 101 sites have been identified along the Maastrichtian–Danian boundary around the globe (Claeys et al. 2002). Around the Gulf of Mexico and the Atlantic Ocean the ejecta layer coincides with erosion of Maastrichtian sediments and is overlain by clastic sediments and breccia attributable to seismic and tsunami effects.

A panel of 41 international experts from 33 institutions concluded the evidence for a cause and effect relation between asteroid impact and mass extinction at the

4.7 K–T (Cretaceous-Tertiary Boundary) Mass Extinction

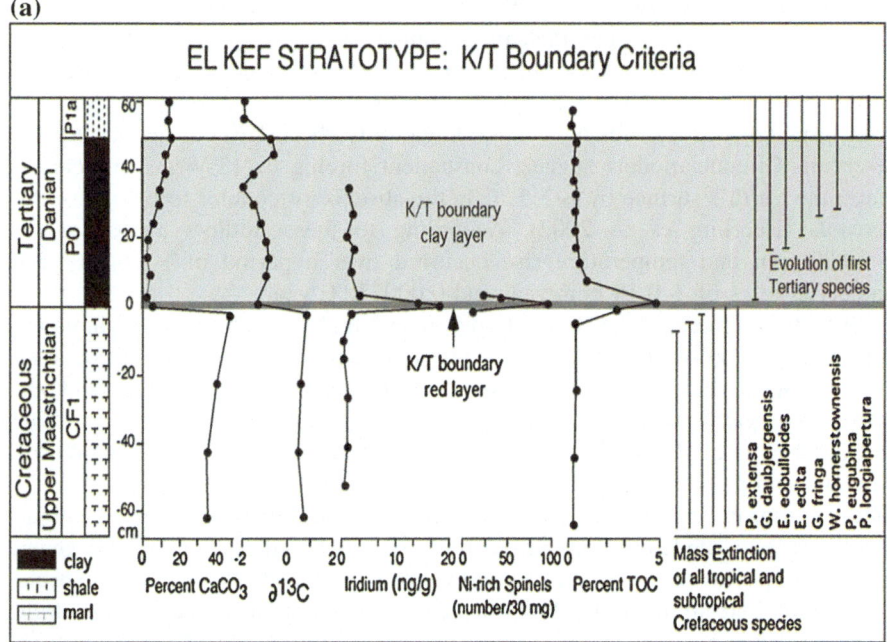

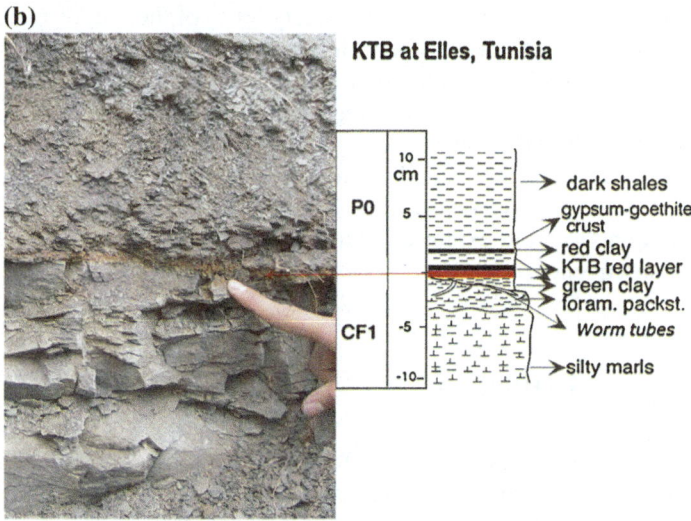

Fig. 4.7 **a** K/T boundary at the El Kef, Tunisia, defined by a dark organic-rich boundary clay with thin red layer at its base enriched in iridium, nickel rich spinels, pyrite and rare clay spherules, a negative $\delta^{13}C$ excursion, high Total Carbon Contents (*TOC*) and low $CaCO_3$, and extinction of tropical and subtropical species (Keller 2005, Fig. 6; Australian Journal of Earth Science, by permission) **b** Photograph showing the K/T boundary at Elles, Tunisia (*Courtesy* G. Keller)

K–T boundary is overwhelming (Schulte et al. 2010). The extinction also coincided with a spike in Deccan Plateau volcanism in India. Stomata leaf pore-based estimates of atmospheric CO_2 during these events indicate an abrupt rise from $\sim$350–500 ppm to at least $\sim$2,300 ppm within about $\sim$10,000 years, consistent with instantaneous transfer of $\sim$4,600 Gigaton Carbon (GtC) to the atmospheric reservoir. Climate models suggest consequent forcing of 12 W/m2 sufficient to warm the Earth's surface by $\sim$7.5 °C in the absence of counter forcing by sulfate aerosols (Beerling et al. 2002). According to these authors a CO_2 rise of $\sim$1,800 ppm and temperature rise occurred over a period of $\sim$10,000 years, namely at rates of $\sim$0.18 ppm/year and 0.00075 °C/year.

Short term effects of the K–T asteroid impact include incineration of large land surfaces, ejection of dust and water vapor, oxidation of atmospheric nitrogen and consequent ozone depletion. Longer term effects included release of CO_2 and other greenhouse gases with consequent warming, ocean acidification and anoxia (McCracken et al. 1994). The K–T mass extinction involved phytoplankton, calcareous nanoplankton, planktonic foraminifera, benthic foraminifera, 54 % of diatoms, marine invertebrates, crustaceans, ostracods, 98 % of tropical colonial corals, 60 % of late Cretaceous Scleractinia coral, echinoderm, and bivalve genera, numerous species of the molluscan and Cephalopoda and all cephalopod species, belemnoids and ammonoids, 35 % of echinoderm genera, rudists (reef-building clams), inoceramids (giant relatives of modern scallops), jawed fishes, cartilaginous fishes (Fig. 4.7a). Survivors included approximately 80 % of the sharks, rays, and skates families. In North America, approximately 57 % of plant species became extinct. All Archaic birds and non-avian dinosaurs became extinct. Cretaceous mammalian lineages, including egg-laying mammals, multi-tuberculates, marsupials and placentals, dryolestoideans, and gondwanatheres survived. Marsupials mostly disappeared from North America and Asian deltatheroidans became extinct. Many of these extinctions constituted proximal instantaneous consequences of the fire ball and asteroid explosion, while distal habitats were affected by more protracted consequences.

4.8 Paleocene-Eocene Extinction

The Paleocene-Eocene thermal maximum (PETM) at $\sim$55.9 Ma involved a release of some $\sim$2,000 billion ton carbon (GtC) as methane, elevating atmospheric CO_2 to near-1,800 ppm at a rate of 0.18 ppm/year and mean temperatures rise of $\sim$5 °C (Zachos et al. 2008; Panchuk et al. 2008; Cui et al. 2011) (Fig. 3.3). Elevated CO_2 led to acidification of ocean water from $\sim$8.2 to $\sim$7.5 pH and the extinction of $\sim$35–50 % of benthic foraminifera during a period of $\sim$1,000 years (Zachos et al. 2008). Other consequences included a global expansion of subtropical dinoflagellate plankton, appearance of modern orders of mammals, including primates, a transient dwarfing of mammalian species, and migration of large mammals from Asia to North America.

4.9 The End-Eocene Freeze

The incidence of an asteroid impact cluster about 35.7–35.6 Ma (Popigai, Siberia—~100 km-diameter; Chesapeake Bay, off-shore Virginia—85 km-diameter; Mount Ashmore, Timor Sea—>50 km-diameter; the related North American strewn tektite field) and the abrupt decline in temperatures about ~33.7–33.5 Ma (Fig. 3.4) can be expected to have triggered major environmental and biotic transformations. Abrupt cooling (Pearson et al. 2009) was associated with elimination of warm-water planktonic species (Keller 1986, 2005). Alvarez (2003) and Poag (1997) suggested impact-related extinctions during the late Eocene and the E-O transition, whereas Keller (2005) ruled out impact-triggered extinction. Isotopic $\delta^{13}C$ and $\delta^{18}O$ studies of late Eocene Iridium-rich ejecta layers at Massignano, Italy, indicate increase in temperature and in organic matter associated with the impacts, possibly reflecting release of methane hydrates by impact excavation (Monechi et al. 2000; Bodiselitsch et al. 2004).

4.10 Carbon and Oxygen Isotopes and Mass Extinctions

Phanerozoic mass extinctions marked by carbon and oxygen isotopic anomalies include the End-Ordovician (Marshall 1992; Marshall et al. 1997; Brenchley et al. 2003), Late Devonian (Stephen and Sumner 2002), Permian-Triassic boundary (Joachimski et al. 2012), Late Triassic (Korte et al. 2009) and K–T boundary (Maruoka et al. 2007). Changes in the carbon cycle recorded by total organic carbon (TOC) and stable carbon isotope ratios ($\delta^{13}C_{carb}$ and $\delta^{13}C_{org}$)[1] constitute sensitive fingerprints of mass burial of organic matter derived from marine organisms, fallout from forest fires, or increased biological productivity. Variations in oxygen isotope ratios ($\delta^{18}O$)[2] reflect changes in ice volumes and salinity. Marked changes in these parameters accompany major volcanic eruptions, asteroid impacts (Maruoka et al. 2007) and methane release (Zachos et al. 2008) (Fig. 3.3).

Positive excursions in both $\delta^{18}O$ and $\delta^{13}Ccarb$ at the end-Ordovician signify parallel decrease in temperature and in biological productivity at the onset of 443.4 ± 1.5 Ma Ashgill/Hirnantian glaciation and extinction event (Marshall 1992; Marshall et al. 1997; Brenchley et al. 2003). A global nature of the glaciation is indicated from the widespread positive carbon isotope, including $\delta^{13}Ccarb$ and $\delta^{13}Corg$, and oxygen isotope shifts measured from Brachiopod shells over a

---

[1] Preferential retention in plants and organisms of ^{12}C relative to ^{13}C measured on well preserved fossils and matrix carbonate allows identification of biological productivity correlated with decreased $\delta^{13}C$ in both organic ($\delta^{13}C_{org}$) and inorganic ($\delta^{13}C_{carb}$) fractions.

[2] Fractionation of oxygen isotopes on evaporation and condensation favors preferential concentration in the air of the light isotope ^{16}O relative to the heavier isotope ^{18}O, leading to heavier $\delta^{18}O$ ratios in water with lower temperature, an index recorded in fossils.

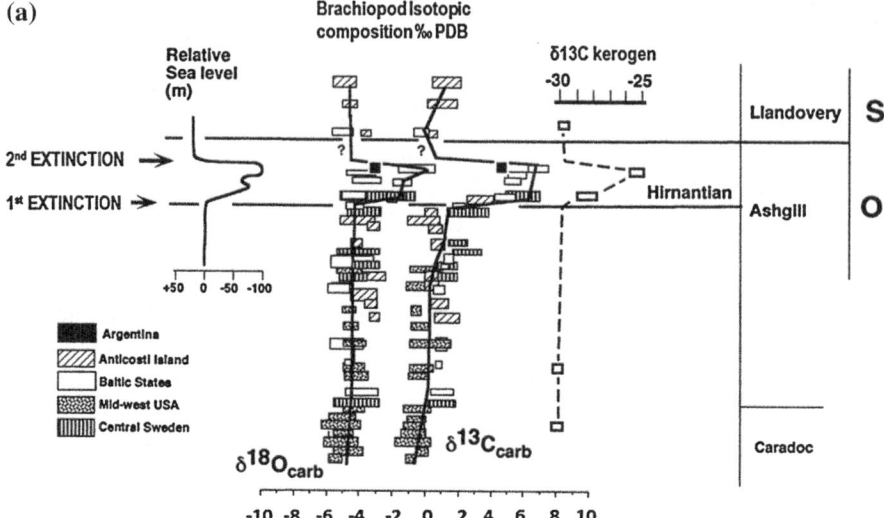

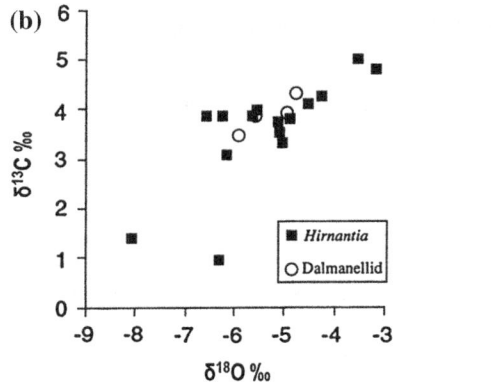

Fig. 4.8 a Summary diagram showing the relationships between biological, bathymetric and stable isotopic changes in the Late Ordovician. The marked positive carbonate isotopic excursion in the early Hirnantian is paralleled by a shift in the isotopic composition of organic carbon. Data from Argentina demonstrate elevated $\delta^{13}C$ in brachiopods from high palaeolatitudes but the oxygen values should be regarded as a minimum because even the least-altered sample showed signs of diagenetic alteration (Marshall et al. 1997, Fig. 9; Elsevier, by permission) **b** The relations between organic matter-related $\delta^{13}C$ and temperature related $\delta^{18}O$ stable isotope proxies in brachiopod samples from the La Pola and Don Braulio sections (Marshall et al. 1997, Fig. 5; Elsevier, by permission)

wide range of paleo-latitudes. Upper Ordovician cores from Estonia and Latvia record a $\delta^{13}Ccarb$ shift of up to 6 ‰ and similar profiles were measured in Nevada, suggesting a global chronostratigraphic signal (Brenchley et al. 2003) (Fig. 4.8a). Positive correlation between $\delta^{13}C$ and $\delta^{18}O$ militates for genetic relations between cooling and decreased biological activity (Fig. 4.8b).

Upper Devonian isotopic excursions were studied at the Fitzroy reef complex of the Canning Basin, Western Australia, where well-exposed continuous Frasnian-Fammenian sequences record interactions between sea level changes, sediment supply, ocean chemistry, and paleoecology (Stephens and Sumner 2002). The Frasnian-Fammenian transition correlates with positive $\delta^{13}C$ shifts, consistent with similar intervals of the Kellwasser horizons in Europe (Fig. 4.9a). By analogy to the end-Ordovician, a positive correlation pertains between $\delta^{13}C$ and $\delta^{18}O$ relations (Fig. 4.9b), suggesting a decline in biological productivity with lower temperatures (Stephens and Sumner 2002). Likely connections between the Frasnian-Fammenian ($\sim$374 Ma) and End-Devonian ($\sim$359 Ma) mass extinctions (Fig. 4.3) and asteroid impacts (Woodleigh, Western Australia $\sim$359 ± 4 Ma; Siljan, Sweden, 376.8 ± 1.7 Ma; Charlevoix, Quebec, 342 ± 15; Alamo, New Mexico, $\sim$360 Ma) (Glikson 2005; Keller 2005) remain to be established once the relations between the stratigraphic context of fallout/ejecta units of these impact are studied.

A marked decline of about 4–7 ‰ $\delta^{13}C$ values across the Permian–Triassic boundary (251 Ma) represents a sharp decline in biological productivity (Figs. 4.4, 4.10), signifying a world-wide mass deposition of extinct biota (Ward 2007; Korte and Kozur 2010). The latter authors defined an overall $\delta^{13}C$ plot displaying a gradual 4 to 7 ‰ decline of $\delta^{13}C$ values lasting over about 0.5 million years, beginning in the Changhsingian (253.8 ± 0.7 Ma) and reaching a first minimum at the P–T boundary (251 ± 0.4 Ma) (Fig. 4.10). The trend is interrupted by two short-term positive excursions following which a decline in $\delta^{13}C$ values continues. Korte and Kozur (2010) interpret these variations in terms of (1) direct and indirect effects of volcanism of the Siberian Traps; (2) anoxic deep waters occasionally reaching very shallow sea levels. The authors question an abrupt release of isotopically light methane from sediments or permafrost soils as a source for the negative carbon-isotope trend. With near to 80 % mass extinction of genera (Keller 2005), a prolonged recovery of the biosphere followed over a period of near to 5 million years (Erwin 2006; Korte and Kozur 2010).

Oxygen isotope ratios measured on phosphate-bound oxygen in conodont apatite from South China decrease by 2 ‰ in the latest Permian, translating into low-latitude surface water warming of 8 °C (Joachimski et al. 2012). The oxygen excursion coincides with a negative $\delta^{13}C_{carb}$ shift, suggesting CO_2 emission by Siberian Trap volcanism constituted a factor driving warming. Temperature rise commenced immediately prior to the main extinction phase, with maximum temperatures documented in the latest Permian at Meishan (bed 27), coinciding with the main pulse of extinction and the collapse of marine and terrestrial ecosystems. Prolonged warming through the Early Triassic may have played a major role in the delayed recovery in the aftermath of the Permian–Triassic crisis.

The end-Triassic extinction at 201.4 Ma, associated with opening of the Central Atlantic Magmatic Province CAMP), is marked by large negative carbon isotope excursion, including a transient increase in CO_2. Carbon isotopic anomalies of leaf wax derived lipids (n-alkanes), wood, and total organic carbon from lacustrine sediments intercalated with CAMP volcanics in eastern North America are similar

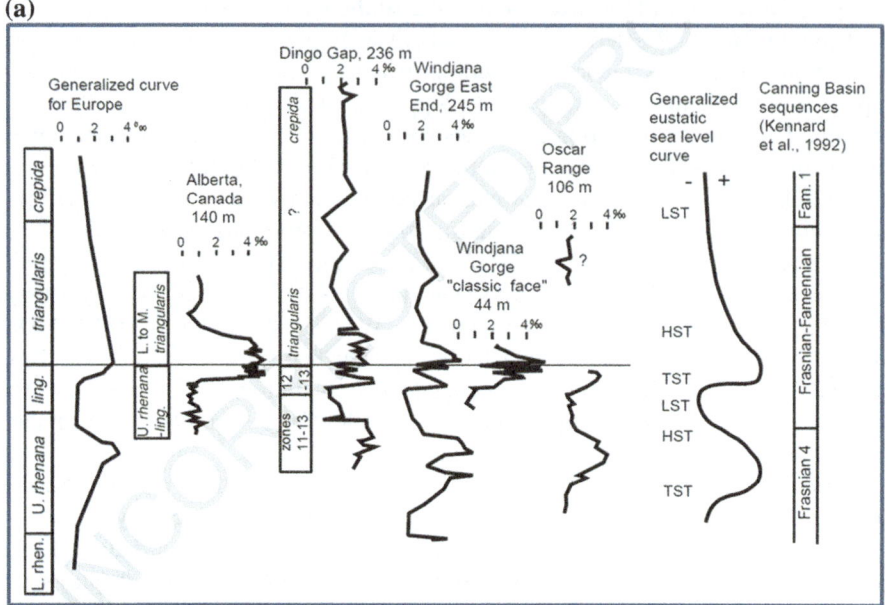

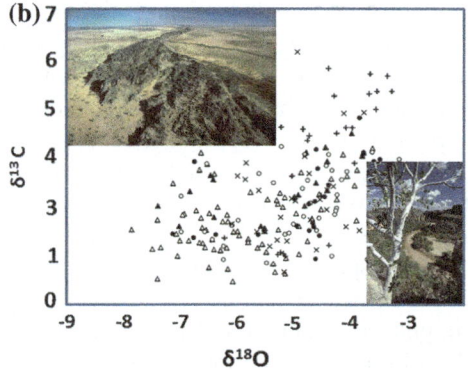

Fig. 4.9 a δ^{13}C curves from the Canning Basin compared to generalized carbon isotope curve from Europe and carbon isotope curve from Canada. The δ^{13}C curve from Europe is plotted against time. The δ^{13}C curves from Australia and Canada are plotted against thickness and adjusted so conodont dates correlate with the European section. Gap in Oscar Range curve is due to unconformity. The generalized eustatic sea level curve is interpreted from Canning Basin and European sections and agrees with Hallam and Wignall's (1997) interpretation (Stephen and Sumner 2002, Fig. 8; Elsevier, by permission) **b** Carbon and oxygen isotopic values of samples used for δ^{13}C correlations (Stephen and Sumner 2002, Fig. 7; Elsevier, by permission). Inset photographs: *top left* Napier Range, Kimberley (*courtesy* Reg Morrison); *bottom right* Windgina Gorge, Kimberley (*Courtesy* Reg Morrison)

Fig. 4.10 General carbon-isotope trend across the P–T boundary constructed from the stratigraphically well-defined $\delta^{13}C$ data from Meishan, Guryul Ravine, Abadeh, Shahreza, Pufels/Bula/Bulla and the Gartnerkofel core. *1* C. changxingensis–C. deflecta Zone, *2* C. zhangi Zone, *3* C. iranica Zone, *4* C. hauschkei Zone, *5* C. meishanensis–H. praeparvus Zone (5 + 6 H. praeparvus Zone for shallow-water without Clarkina such as the Southern Alps), *6* M. ultima–S. *7* H. parvus Zone = Triassic part of C. zhejiangensis (for South Chinese intra-platform basins), *8* I. isarcica Zone (Korte and Kozur 2010, Figs. 1 and 5; Elsevier, by permission)

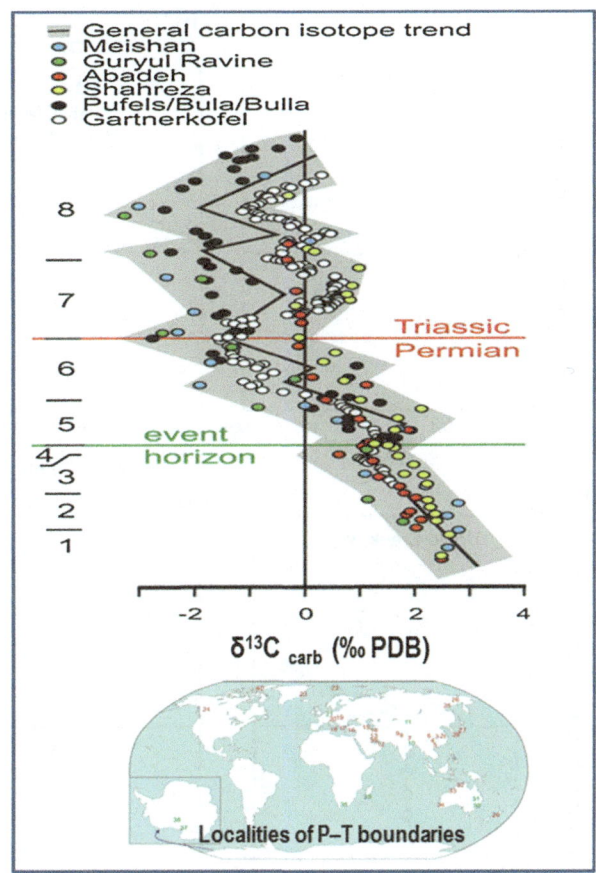

to anomalies in sections across the Atlantic, suggesting synchronous onset of the extinction (Whiteside et al. 2010). The onset of the anomalies precedes the oldest basalts in eastern North America but are simultaneous with the eruption of the oldest flows in Morocco, signifying a CO_2 super greenhouse and marine acidification crisis. Bachan et al. (2012) report $\delta^{13}C$ anomalies from six newly sampled sections in Italy displaying sharp negative $\delta^{13}Ccarb$ excursions coincident with the disappearance of the Triassic fauna and two overlying positive excursions. No negative $\delta^{13}Corg$ were recorded, suggesting diagenetic alteration of organic matter. The data suggest perturbation of the global carbon cycle persisting for substantial length of geologic time following the extinction.

The most extensively studied asteroid impact boundary of the 65 Ma K–T event has yielded definitive carbon, oxygen and sulphur isotopic values diagnostic of the geological, geochemical and biological effects of large-scale impact.

Detailed studies of TOC and $\delta^{13}C$ anomalies across the K–T boundary in freshwater floodplains and swamp environments of Montana and Wyoming by

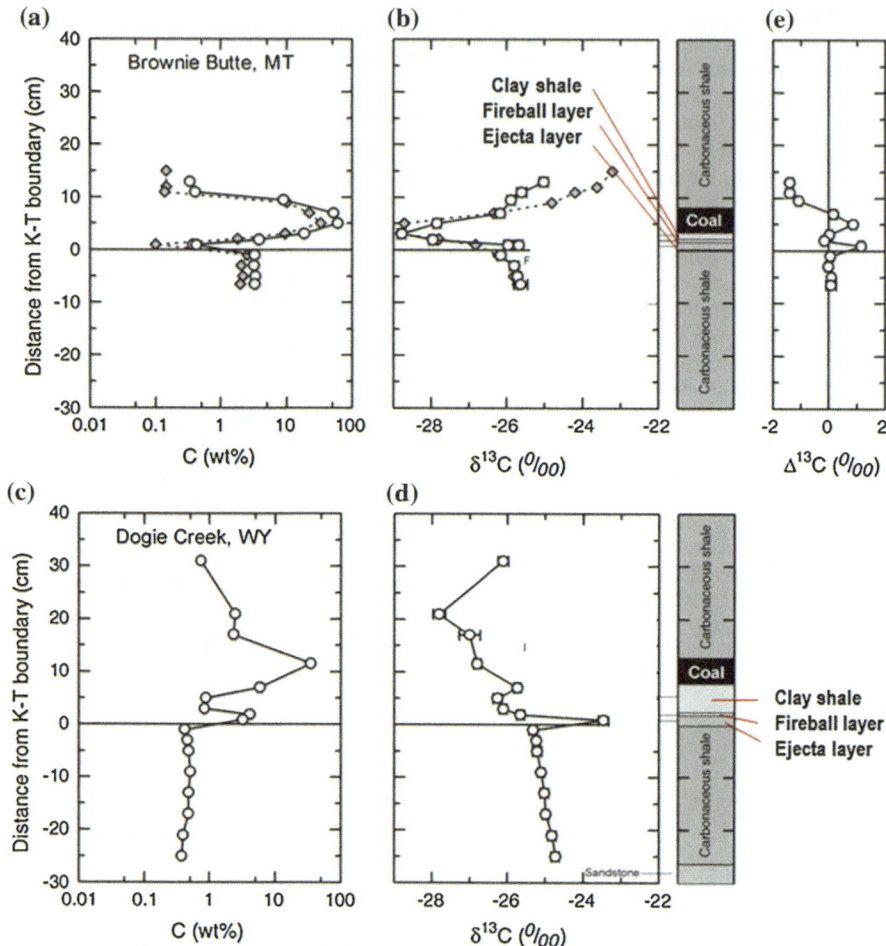

Fig. 4.11 Chemostratigraphic profiles for **a** organic carbon concentrations and **b** $\delta^{13}C$ values of bulk organic carbon for the Brownie Butte samples and for **c** organic concentrations and **d** $\delta^{13}C$ values of bulk organic carbon for the Dogie Creek samples. *Open circles* and *gray diamonds* represent data obtained by Maruoka et al. (2007) and those obtained by Gardner and Gilmour (2002), respectively. (Maruoka et al. 2007, Fig. 4; Elsevier, by permission)

Maruoka et al. (2007) reveal a marked decrease of $\delta^{13}C$ values by 2.6 ‰ (from −26.15 ‰ to −28.78 ‰) at Brownie Butte (Fig. 4.11), similar to the trend in carbonate at marine K–T sites. The $\delta^{13}C_{org}$ values are thought to reflect variations in carbon isotopes in atmospheric CO_2 in equilibrium with the ocean surface water. Other factors include enhanced contribution of organic matter derived from algae in a high-productivity environment due to nitrogen fertilization and/or eutrophication induced by sulfide. The authors suggest the high productivity recorded in the K–T boundary clays imply that, by contrast to marine environments, freshwater environments recovered rapidly from the effects of impact. At a second K–T impact boundary site of Dogie Creek a positive shift of $\delta^{13}C$org is observed (Fig. 4.11) similar to other continental sites in North America. Variations between the sections suggest the effects of local environments such as anoxia and reactions with sulphide and sulphate related to acid rain effects of the impact.

Kaiho et al. (1999), reporting 36 isotopic analyses of K–T boundary samples from Caravaca, Spain, observe a rapid reduction in the gradient between $\delta^{13}C$ values in fine fraction carbonate and benthic foraminiferal calcite in sediments immediately above the K/T boundary, implying an abrupt extinction of pelagic organisms leading to a significant reduction in the flux of organic carbon to the seafloor. Variations in sulfur isotope ratios at Caravaca, Japan and New Zealand imply a rapid decrease in oxygen coincident with the $\delta^{13}C$ shift. A three-fold increase in the kaolinite/illite ratio and a 1.2 ‰ decrease in $\delta^{18}O$carb recorded in the basal 0.1–2 cm of Danian sediments indicate atmospheric and marine warming up to 3,000 years following the $\delta^{13}C$ event. Recovery commenced some 13,000 years kyr following the K–T event.

References

Alvarez LW, Alvarez W, Asaro F, Michel HV (1980) Extra-terrestrial Cause for the Cretaceous-TertiaryExtinction: Experimental results and theoretical interpretation. Science 208:1095–11086
Alvarez W (2003) Comparing the evidence relevant to impacts and flood basalts at times of major mass extinctions. Astrobiology 3:153–161
Bachan A, Van de Schootbrugge B, Fiebig J, McRoberts C, Ciarapica G, Payne J (2012) Carbon cycle dynamics following the end-triassic mass extinction: constraints from paired $\delta^{13}C$carb and $\delta^{13}C$org. Geochem Geophys Geosyst. doi:10.1029/2012GC004150
Balter V et al (2008) Record of climate-driven morphological changes in 376 Ma Devonian fossils. Geology 36:907
Beerling DJ (2002a) CO_2 and the end-Triassic mass extinction. Nature 415:386–387
Beerling DJ (2002b) Low atmospheric CO_2 levels during the Permo-Carboniferous glaciation inferred from fossil lycopsids. Proc Nat Acad Sci 99:12567–12571
Beerling DJ, Osborne CP, Chaloner WG (2001) Evolution of leaf-form in land plants linked to atmospheric CO2 decline in the Late Palaeozoic era. Nature 410:352–354
Beerling DJ, Lomax BH, Royer DL, Upchurch GR, Kump LR (2002) An atmospheric pCO_2 reconstruction across the Cretaceous-Tertiary boundary from leaf mega fossils. Proc Nat Acad Sci 99:7836–7840
Berner RA (2005) The carbon and sulfur cycles and atmospheric oxygen from middle Permian to middle Triassic. Geochim Cosmochim Acta 69:3211–3217
Bodiselitsch B, Montanari A, Koeberl C, Coccioni R (2004) Delayed climate cooling in the Late Eocene caused by multiple impacts: high-resolution geochemical studies at Massignano, Italy. Earth Planet Sci Lett 223:283–302
Brenchley PJ, Carden GA, Hints L, Kaljo D, Marshall JD, Martma T, Meidla T, Nõlvak J (2003) High-resolution isotope stratigraphy of Late Ordovician sequences: constraints on the timing of bio-events and environmental changes associated with mass extinction and glaciation. Geol Soc of Am Bull 115:89–104

Calver CR (2000) Isotope stratigraphy of the Ediacarian (Neoproterozoic III) of the Adelaide Rift Complex, South Australia, and the overprint of water column stratification. Precamb Res 100:121–150

Claeys P, Kiessling W, Alvarez W (2002) Distribution of Chicxulub ejecta at the Cretaceous-Tertiary boundary in Koeberl C and MacLeod KG eds Catastrophic Events and Mass Extinctions: Impacts and Beyond. Geol Soc Am Spec Pap 356:55–68

Courtillot VE, Rennes PR (2003) On the ages of flood basalt events. CR Geosci 335:113–140

Cui Y, Kump LR, Ridgwell AJ, Charles AJ, Junium CK, Diefendorf A.F, Freeman KH, Urban NM,Harding IC (2011) Slow release of fossil carbon during the Palaeocene–Eocene Thermal Maximum.Nature Geosci 4:481–485

Erwin DH (2006) Extinction: how life on earth nearly ended 250 million years-ago. Princeton Univ Press, Princeton and Oxford, p 296

Frakes LA, Francis JE, Syktus JI (1992) Climate modes of the phanerozoic. Cambridge University Press, Cambridge

French BM (1998) Traces of Catastrophe. Lunar Planetary Institute 954:120

Gardner AF, Gilmour I (2002) An organic geochemical investigation of terrestrial Cretaceous–Tertiary boundary successions from Brownie Butte, Montana, and the Raton Basin, New Mexico. In: Koeberl C, MacLeod KG (eds) Catastrophic events and mass extinctions: impacts and beyond. Geol Soc Am Spec Pap 356:351–362

Glass LM, Phillips D (2006) The Kalkarindji continental flood basalt province: A new Cambrian large igneous province in Australia with possible links to faunal extinctions. Geology 34:461–464

Glikson AY (2005) Asteroid/comet impact clusters, flood basalts and mass extinctions: significance of isotopic age overlaps. Earth Planet Sci Lett 236:933–937

Glikson AY (2013) The asteroid impact connection of planetary evolution. Springer, Dordrecht, p 149

Glikson AY, Uysal IT, Fitz Gerald JD, Saygin E (2013a) Geophysical anomalies and quartz microstructures, Eastern Warburton Basin, North-east South Australia: Tectonic or impact shock metamorphic origin? Tectonophysics 589:57–76

Glikson AY, Uysal IT (2013b) Geophysical and structural criteria for the identification of buried impact structures, with reference to Australia. Earth Sci Rev 125:114–122

Gostin VA, Zbik M (1999) Petrology and microstructure of distal impact ejecta from the Flinders Ranges Australia. Metor Planet Sci 34:587–592

Gostin VA, Haines PW, Jenkins RJF, Compston W, Williams IS (1986) Impact ejecta horizon within late Precambrian shale, Adelaide Geosyncline, South Australia. Science 233:198–200

Gradstein FM, Ogg JG (2004) Geologic time scale 2004—why, how, and where next. Lethaia 37:175–181

Grasby SE, Sanei H, Beauchamp B (2011) Catastrophic dispersion of coal fly ash into oceans during the latest Permian extinction. Nat Geosci 4:104–107

Grey K (2005) Ediacaran palynology of Australia, vol 31. Association of Australasian Palaeontologists Mem, Canberra, p 439

Grey K, Walter MR, Calver CR (2003) Neoproterozoic biotic diversification: snowball earth or aftermath of the Acraman impact? Geology 5:459–462

Hallam A, Wignall PB (1997) Mass extinctions and their aftermath. Oxford University Press, Oxford

Hames W, McHone JG, Renne P, Ruppel C (2003) The Central Atlantic Magmatic Province: insights from fragments of Pangea. Geophys Monog Series 136:267

Joachimski MM, Xulong L, Shen S, Jiang H, Chen B, Sun Y (2012) Climate warming in the latest Permian and the Permian–Triassic mass extinction. Geology 40:195–198

Jourdan F, Marzoli A, Bertrand HS, Cirilli S, Tanner LH, Kontak DJ, McHone G, Renne PR, Bellieni G (2009) 40Ar/39Ar ages of CAMP in North America: implications for the Triassic–Jurassic boundary and the 40 K decay constant bias. Lithos 110:167–180

Kaiho KY, Kajiwara K, Tazaki M, Ueshima N, Takeda H, Kawahata T, Arinobu R, Ishiwatari A, Hirai MA (1999) Oceanic primary productivity and dissolved oxygen levels at the Cretaceous/

Tertiary Boundary: Their decrease, subsequent warming, and recovery. Paleoceanography 14:511–524

Kamo SL, Czamanske GK, Amelin Y, Fedorenko VA, Davis DW, Trofmov VR (2003) Rapid eruption of Siberian flood-volcanic rocks and evidence for coincidence with the Permian-Triassic boundary and mass extinction at 251 Ma. Earth Planet Sci Lett 214:75–91

Keller G (1986) Stepwise mass extinctions and impact events; late Eocene to early Oligocene. Mar Micropaleontol 10:267–293

Keller G (2005) Impacts volcanism and mass extinction: random coincidence or cause and effect? Aust J Earth Sci 52:725–757

Korte C, Hesselbo SP, Jenkyns HC, Rickaby REM (2009) Palaeo-environmental significance of carbon- and oxygen-isotope stratigraphy of marine Triassic–Jurassic boundary sections in SW Britain. J Geol Soc London 166:431–445

Korte C (2010) Kozur HW (2010) Carbon-isotope stratigraphy across the Permian–Triassic boundary: a review. J Asian Earth Sci 39:215–235

Kump LR, Arthur MA, Patzkowsky ME, Gibbs MT, Pinkus DS, Sheenan PM (1999) A weathering hypothesis for glaciation at high atmospheric pCO_2 during the Late Ordovician. Palaeoclimatol Palaeogeogr Palaeoecol 152:173–187

Marshall JD (1992) Climatic and oceanographic isotopic signals from the carbonate rock record and their preservation. Geol Mag 129:143–160

Marshall JD, Brenchley PJ, Mason P, Wolff GA, Astini RA, Hints L, Meidla T (1997) Global carbon isotopic events associated with mass extinction and glaciation in the Late Ordovician. Palaeo, Palaeoclim, Palaeoecol 132:195–210

Maruoka T, Koeberl C, Bohor BF (2007) Carbon isotopic compositions of organic matter across continental Cretaceous-Tertiary (K-T) boundary sections: Implications for paleo environment after the K-T impact event. Earth Planet Sci Lett 253:226–238

McCracken MC, Convey C, Thompson SL, Weissman PR (1994) Global climatic effects of atmospheric dust from an asteroid or comet impact on Earth, Glob Planet Change 9:263–273

McElwain JC, Punyasena SW (2007) Mass extinction events and the plant fossil record. Trends Ecol Evol 22:549–557

McElwain JC, Beerling DJ, Woodward FI (1999) Fossil plants and global warming at the Triassic-Jurassic boundary. Science 285:1386–1390

McGhee GR (1996) The late Devonian mass extinction. Columbia University Press, New York

Monechi S, Buccianti A, Gardin S (2000) Biotic signals from nannoflora across the iridium anomaly in the upper Eocene of the Massignano section: evidence from statistical analysis. Mar Micropaleontol 39:219–237

Mora CI, Driese SG, Colarusso LA (1996) Middle to late Paleozoic atmospheric CO_2 levels from soil carbonate and organic matter. Science 271:1105–1107

Olsen PE, Sues HD (1986) Correlation of continental late Triassic and early Jurassic sediments and patterns of the Triassic: Jurassic tetrapod transition. In: Padian K (ed) The beginning of the age of Dinosaurs. Cambridge University Press, Cambridge, p 321–351

Panchuk K, Ridgwell A, Kump LR (2008) Sedimentary response to Paleocene-Eocene thermal maximum carbon release: a model-data comparison. Geology 36:315–318

Pearson PN, Foster GL, Wade BS (2009) Atmospheric carbon dioxide through the Eocene–Oligocene climate transition. Nature 461:1110–1113

Poag CW (1997) Roadblocks on the kill curve: testing the Raup hypothesis. Palaios 12:582–590

Pope KO, Baines KH, Ocampo AC, Ivanov BA (1997) Energy volatile production and climatic effects of the Chicxulub Cretaceous/Tertiary impact. J Geophys Res 102:21645–21664

Racki G (2003) End-Permian mass extinction: oceanographic consequences of double catastrophic volcanism. Lethaia 36:171–173

Renne PR, Zhang Z, Richards MA, Black MT, Basu AR (1995) Synchrony and causal relations between Permian: Triassic boundary crises and Siberian flood volcanism. Science 269:1413–1416

Ross CA, Ross RP (1995) Permian sequence stratigraphy. In: Scholle PA et al. (eds) The Permian of northern Pangea, vol 1. Springer, Berlin, p 98–123

Rothwell GW, Scheckler SE, Gillespie WH (1989) Elkinsia gen nov a Late Devonian gymnosperm with cupulate ovules. Bot Gaz 150:170–189

Royer DL (2006) CO2-forced climate thresholds during the Phanerozoic. Geochim Cosmochim Acta 70:5665–5675

Schulte P, Alegret L, Arenillas I, Arz JA, Barton PJ, Bown PR, Bralower TJ, Christeson GL et al (2010) The Chicxulub Asteroid impact and mass extinction at the Cretaceous-Paleogene Boundary. Science 327(5970):1214–1218

Sepkoski JJ (1996) Patterns of Phanerozoic extinction: a perspective from global data bases. In: Walliser OH (ed) Global events and event stratigraphy. Springer, Berlin, p 35–52

Stephens NP, Sumner DY (2002) Late Devonian carbon isotope stratigraphy and sea level fluctuations, Canning Basin, Western Australia. Palaeo 3024:1–17

Tohver E et al (2012) Geochronological constraints on the age of a Permo-Triassic impact event: U-Pb and 40Ar/39Ar results for the 40 km Araguainha structure of central Brazil. Geochim et Cosmochim Acta 86:214–227

Twitchett RJ, Looy CV, Morante R, Visscher H, Wignall PB (2001) Rapid and synchronous collapse of marine and terrestrial ecosystems during the end-Permian crisis. Geology 29:351–354

Walter MR, Veevers JJ, Calver CR, Gorjan, Hill AC (2000) Dating the 840–544 Ma Neoproterozoic interval by isotopes of strontium, carbon, and sulfur in seawater, and some interpretative models: Precamb Res 100:371–433

Ward PD (2007) Under a green sky: global warming, the mass extinctions of the past, and what they can tell us about our future. Harper Collins, New York, p 242

Whiteside JH, Olsen PE, Eglinton T, Brookfield ME, Sambrotto RN (2010) Compound-specific carbon isotopes from Earth's largest flood basalt eruptions directly linked to the end-Triassic mass extinction. Proc Nat Acad Sci, pnas.1001706107

Wignall PB (2001) Large igneous provinces and mass extinctions. Earth Sci Rev 53:1–33

Wignall PB, Twitchett RJ (1996) Oceanic anoxia and the end Permian mass extinction. Science 272:1155–1158

Williams GE, Gostin VA (2005) The Acraman—Bunyeroo impact event (Ediacaran) South Australia and environmental consequences: 25 years on. Aust J Earth Sci 52:607–620

Williams GE, Schmidt PW, Boyd DM (1996) Magnetic signature and morphology of the Acraman impact structure South Australia. Aust Geol Surv Org J Aust Geol Geophys 16:431–442

Zachos J, Dickens GR, Zeebe RE (2008) An early Cenozoic perspective on greenhouse warming and carbon-cycle dynamics. Nature 451:279–283

Part III
Homo's Fire Blueprint

I want to know God's thoughts—the rest are details
Albert Einstein

Chapter 5
A Flammable Biosphere

Abstract The advent of plants on land surfaces since about 420 million years-ago created an interface between carbon-rich organic layers and an oxygen-rich atmosphere, leading to recurrent fires triggered by lightning, volcanic eruptions, high-temperature combustion of peat and, finally, ignition by humans, constituting the blueprint for the Anthropocene. For a species to be able to control ignition and energy output, leading to increase in entropy in nature higher by orders of magnitude than its own physical energy outputs, the species would need to be perfectly wise and responsible. No species can achieve such levels.

The emergence of land plants in the late Silurian ∼420 Ma, the earliest being vascular plants (Cooksonia, Baragwanathia) and later Cycads and Ginkgo in the Permian (299–251 Ma) (Figs. 1.2, 5.1), combined with the rise in photosynthetic oxygen above 13 % (Fig. 2.2), with consequent juxtaposition of carbon-rich land surfaces and rising atmospheric oxygen, have set the stage for extensive fires. Fires became an integral part of the atmosphere/land carbon and oxygen cycles and have led to development of fire-adapted (pyrophyte) plants, enhancing the distribution of seeds and control of parasites. With the exception of anaerobic chemo-bacteria which metabolize sulphur, carbon and metals, photosynthesis has become the basis for the land-based food chain. oxidation reactions through fire and by plant-consuming organisms enhance degradation and entropy. The decay of plants coupled with charcoal from fires coated many parts of the land with carbon, located in cellulose of trees and grasses, in soils and marshes, in methane hydrate and methane clathrate deposits in bogs, sediments and permafrost.

Charcoal, a proxy for fire, occurs in the fossil record from the Late Silurian ∼420 Ma. Scott and Glasspool (2006) document Silurian through to end-Permian charcoal deposits with reference to the frequency of Paleozoic fires in relation to atmospheric oxygen concentrations. As atmospheric oxygen levels rose from ∼13 % in the Late Devonian to ∼30 % in the Late Permian, fires progressively occur in an increasing diversity of ecosystems. Late Silurian to early Devonian charcoal indicates the burning of diminutive rhyniophytoid vegetation. There is an apparent paucity of charcoal in the Middle to Late Devonian coinciding with low atmospheric oxygen (Fig. 2.2). Fires become widespread during

Fig. 5.1 Fossil Palaeozoic plants. **a** Baragwanathia longifolia—Silurian; **b** Baragwanathia longifolia—Silurian, Yea in Victoria; **c** Rhacopteris ovate—Carboniferous; **d** Rhacopterid with partly divided leaves—Carboniferous. (From 'Greening of Gondwana' 1986, Reed Books, French's Forest; photographs by Jim Frazier; *Courtesy* Mary White)

the Early Mississippian (Lower Carboniferous) and in the Middle Mississippian. During the Pennsylvanian (Upper Carboniferous) oxygen rose toward levels above 30 %. Charcoal is recorded in upland settings and is important in many Permian mire settings, suggesting the burning of even moist vegetation. The decline of oxygen levels through much of the Mesozoic (250–65 Ma) to below 15 % (Fig. 2.2) and its gradual resurgence through the late Mesozoic and Cenozoic limited the effect of fire.

Atmospheric CO_2 levels are buffered by the oceans ($\sim$37,000 GtC), which contain about x46 times the atmospheric CO_2 inventory ($\sim$800 GtC). The solubility of CO_2 in water decreases with higher temperature and salinity and the transformation of the $CO_3^{[-2]}$ ion to carbonic acid ($HCO_3^{[-1]}$) retards the growth of calcifying organisms, including corals and plankton. Plants and animals work in opposite directions of the entropy scale, where plants synthesize complex organic compounds from CO_2 and water, producing oxygen, whereas animals burn oxygen and expel CO_2. Disturbances in the carbon and oxygen balance occur when changes occur in the extent of photosynthetic processes, CO_2 solubility in the oceans, burial of carbon in carbonate and organic remains of plants and oxidation of carbon through fire and combustion.

Prior to the mastery of fire by Hominins, wildfires were ignited by lightning, incandescent fallout from volcanic eruptions, meteorite impacts and combustion of peat. Consuming vast quantities of biomass, fires have an essential role in terrestrial biogeochemical cycles (Belcher et al. 2010), including consequences for

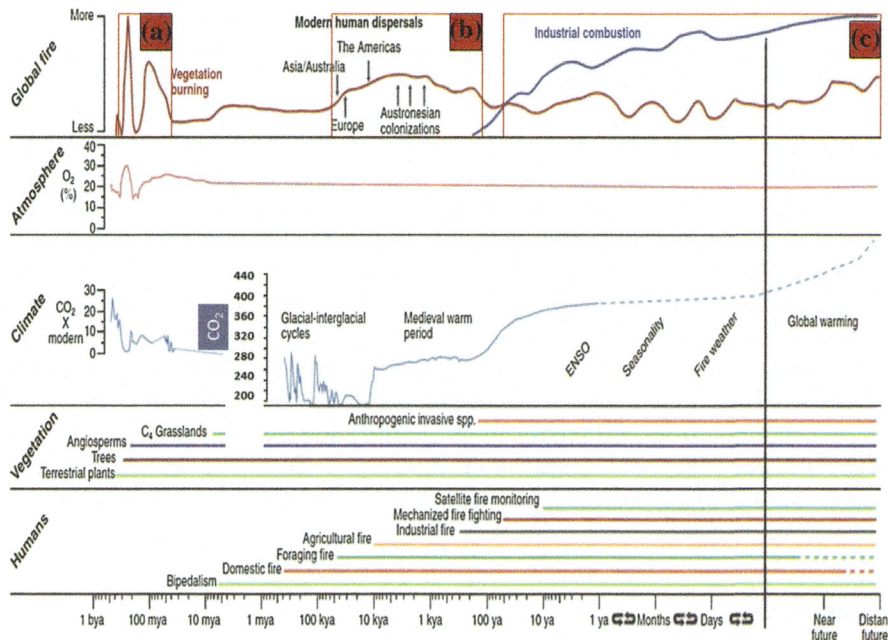

Fig. 5.2 Qualitative scheme of global fire activity through time, based on pre-Quaternary distribution of charcoal, Quaternary and Holocene charcoal records, and modern satellite observations, in relation to the percentage of atmospheric O_2 content, CO_2 (in parts per million), appearance of vegetation types, and presence of hominins. *Dotted lines* indicate periods of uncertainty. **a**—Mesozoic and Palaeozoic flammability peaks; **b**—historic and pre-historic flammability period, in part related to human-lit fires; **c**—Anthropocene fuel combustion; (Bowman et al. 2009, Fig. 1; American association for advancement of science, by permission)

the oxygen cycle and the evolution of biodiversity over geological timescales. Subsequent burial of carbon in sediments stored the fuel over geological periods, affecting the carbon and oxygen cycle in favor of oxygen. The role of extensive fires during the Paleozoic and Mesozoic (Fig. 5.2) is represented by charcoal remains whose pyrogenetic origin is identified by high optical refractive indices. Experiments and models by Belcher et al. (2010) suggest fire is suppressed below 18.5 % O_2, switched off below 16 % O_2 and enhanced between 19 and 22 % O_2. According to Belcher et al. (2010) fires were important during ~350–300 Ma and 145–65 Ma, intermediate effects during the 299–251, 285–201 and 201–145 Ma and low effects between 250 and 240 Ma. During the Carboniferous-Permian period, when atmospheric oxygen levels reached ~31 % or higher (Beerling and Berner 2000; Berner et al. 2007), instantaneous combustion affected even moist vegetation (Scott and Glasspool 2006; Bowman et al. 2009). Thus, Permian (299–251 Ma) coals may contain charcoal concentrations as high as 70 % (Bowman et al. 2009).

Above a certain level of atmospheric oxygen fires would constrain the development of forests, constituting strong negative feedback against excessive rise of atmospheric oxygen (Watson et al. 1978). Conversely a decline of oxygen reduces the frequency and intensity of fires. The association of fossil charcoal with fossil trees suggests O_2 levels continued to be replenished, whereas the upper oxygen limit of Phanerozoic atmospheres is uncertain. Robinson (1989) pointed to Paleobotanical evidence for a higher frequency of fire-resistant plants during the Permo-Carboniferous, supporting distinctly higher O_2 levels at that time. Model calculations of the interaction between terrestrial ecosystems and the atmosphere by Beerling and Berner (2000) suggest the rise from 21 to 35 % O_2 during the Carboniferous resulted in a decline in organic productivity of about 20 % and a loss of more than 200 GtC (billion ton carbon) in vegetation and soil carbon storage, due to burning in an atmosphere of ~ 300 ppm CO_2. However, in a CO_2-rich atmosphere of ~ 600 ppm carbon fertilization of the soil productivity increases lead to the net sequestration of 117 GtC. In both cases these effects resulted from strong interaction between O_2, CO_2 and climate in the tropics.

References

Beerling DJ, Berner RA (2000) Impact of a permo-carboniferous high O_2 event on the terrestrial carbon cycle. Prof Nat Acad Sci 97:12428–12432
Belcher CM, Yearsley JM, Hadden RM, McElwain JC, Rein G (2010) Baseline intrinsic flammability of Earth's ecosystems estimated from paleo-atmospheric oxygen over the past 350 million years. Proc Nat Academy Science. doi10.1073/pnas.1011974107
Berner RA, Vanderbrook JM, Ward PD (2007) Oxygen and evolution. Science 316:557–558
Bowman DM et al (2009) Fire in the earth system. Science 324:481–484
Robinson JM (1989) Phanerozoic O_2 variation, fire and terrestrial ecology. Palaeogeogr Palaeoclimatol Palaeoecol 75:223–240
Scott AC, Glasspool IJ (2006) The diversification of paleozoic fire systems and fluctuations in atmospheric oxygen concentration. Proc Nat Acad Sci 103:10861–10865
Watson A, Lovelock JE, Margulis L (1978) Methanogenesis, fires and the regulation of atmospheric oxygen. Biosystems 10:293–298

Chapter 6
A Fire Species

Abstract Evolved in relatively sheltered sub-tropical rift valleys, unique among all genera the genus Homo learnt how to ignite and transfer fire and through this to modify extensive land surfaces of Earth, with consequences for the composition of the atmosphere, a process culminating in the Anthropocene and in the Sixth mass extinction of species.

Progressive overall and intermittent cooling through the early Pleistocene, culminating with the ice ages, saw the retreat of tropical rainforests and the opening of savannah, heralding a fundamental shift in the terrestrial habitats, the rise of Hominins in Africa and subsequent migrations of a succession of human species (Klein and Edgar 2002; Klein 2009; Henn et al. 2012). Once humans became hunters in the open savanna, they became affected by the same climate and environment variability factors which affected their prey, as indicated by the extinction/radiation relations of Bovids (Fig. 6.1) (deMenocal 2004).

The genus Homo is defined by its bipedalism, an increase in its cranial volume from Australopithecines (*A. garhi* ~450 cc) (~2.3–2.4 Ma), to Hominines (*H. habilis* ~600 cc (2.4–2.3 Ma); *H. erectus* 850–1,100 cc (~1.8 Ma); *H. heidelbergensis* ~1,100–1,400 cc (~0.6 Ma); *H. neanderthalensis* 1,200–1,700 cc and *H. sapiens* 1,350–1,400 cc) (Groves 1993; Klein and Edgar 2002; Zimmer 2005; McHenry 2009) (Fig. 6.2). The use of tools included the Oldowan stone industry (2.6–1.7 Ma) and the more sophisticated Acheulean stone industry (from ~1.7 Ma). However, the use of tools is not unique to Homo, as it is shared by other biological genera, including Chimpanzee, beavers, birds and insects. Uniquely it is the mastery of fire which distinguishes the genus Homo from other members of the animal kingdom.

Physical criteria used to distinguish Homo from animals do not adequately explain the nature of the genus. Partial bipedalism, including a switch between two and four legged locomotion, is common among mammals, cf. bears, meerkats, lemurs, gibbons, kangaroos, sprinting lizards, birds and their dinosaur ancestors. Homo sapiens' brain mass is lower than that of whales and elephants but the brain to bodyweight ratio is one to two orders of magnitude higher (human brain 1.3–1.4 kg, average body ~70 kg, ratio ~0.02; sperm whales brain ~7.8 kg, body 40,000 kg, ratio 0.0002; elephants brain ~4.8 kg, body ~5,000 kg, ratio ~0.001)

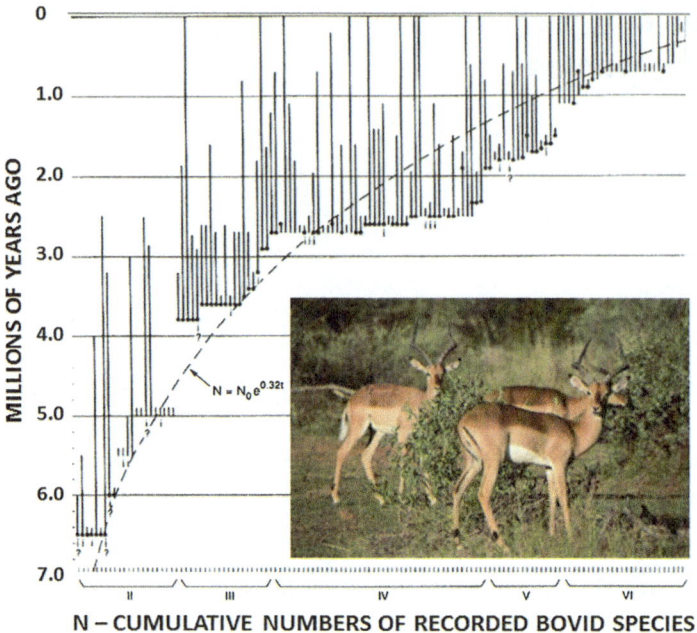

Fig. 6.1 Africa-wide occurrences of fossil bovids (antelope) over the last 7 Ma. Pronounced faunal turnover pulses about ~2.8 and 1.8 Ma were associated with onset of arid-adapted fauna and were more gradual in some areas (deMenocal 2004, Fig. 9; Elsevier, by permission). Photograph of Male Impala (Aepyceros melampus melampus) in: Serengeti Wikipedia Commons; author: Ikiwaner/Wikimedia Commons http://en.wikipedia.org/wiki/File:Serengeti_Impala3.jpg

(Fig. 6.3b). The human brain/body mass ratio is similar to that of mice but lower than that of birds (~0.08). The high neocortex to brain ratio (the so-called 'Dunbar index') of birds has been related to their high sociability and enhanced communications (Dunbar 1996) (Fig. 6.4a). The evolution of hominin, corresponding to the brain weight to body weight ratio (Fig. 6.3a), is represented by the emergence of species from about 7 million years-ago (Sahelanthropus tchadensis—brain case 320–380 cc;) to Homo sapiens neanderthalis (brain case ~1,200–1,500 cc) (Figs. 6.2 and 6.3a)

Numerous organisms use tools and construct articulate structures, examples being the elaborate architecture of termite nests, bee hives, spider webs and beaver dams, and the use of rudimentary tools by some primates, including chimpanzees and orangutans. Examples of sophisticated language among animals include the bee dance, bird songs and the echo sounds of whales and dolphins, possibly not less complex than the languages of original prehistoric humans, if not that of Shakespeare some two million years later. Other features unsurpassed by humans include the visual memory of birds and navigation by ants, birds, fish, whales and insects (Griffin 1992; Narby 2005).

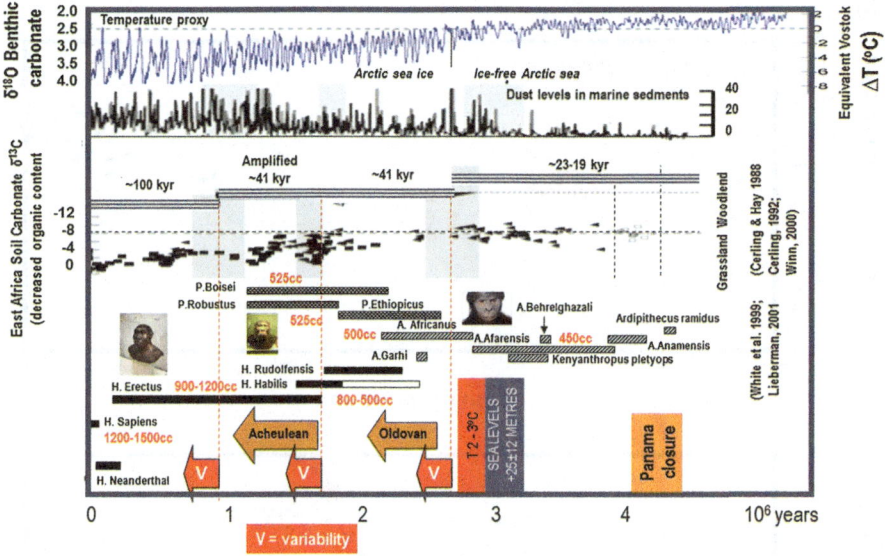

Fig. 6.2 Climate and hominin evolution events during the Pliocene–Pleistocene. Marine paleoclimate records indicate that African climate became progressively more arid after step-like shifts near ~2.8 Ma, and subsequently after ~1.7 and 1.0 Ma, coincident with the onset and intensification of high-latitude glacial cycles. Consequences included changes toward dry-adapted African faunal compositions, including steps in hominin speciation, adaptation, and behavior. Soil carbonate carbon isotopic data from East African hominin fossil localities document the Pliocene–Pleistocene progressive shifts from closed woodland forest (C3-pathway) vegetation to arid-adapted (C4-pathway) savannah grassland vegetation (deMenocal 2004, Fig. 11; added details by the author; Elsevier, with permission)

The appearance of a species which has learnt how to kindle fire (Figs. 6.5, 6.6) meant that, for the first time the flammable carbon-rich biosphere could be ignited by a living organism. The mastery of fire in the mid-Pleistocene about ~2.0–1.0 Ma coincides with accentuation of intermittent glacial conditions associated with the amplification of the 41 kyr-long Milankovic cycles at about 1.8–1.5 Ma, a period of increased climate variability (Figs 3.6–3.11, 6.2). It is likely that, like other major inventions, the mastery of fire was driven by necessity, under acute environmental pressures associated with the descent from warm Pliocene (5.2–2.6 Ma) climate to Pleistocene (2.6–0.01 Ma) (deMenocal 2004), which involved global temperature oscillations of up to ± 5 °C (Figs 3.6b–3.10, 6.2).

Examples of the effects of climate on early human evolution come from East African rift valleys, where climates were controlled by major tectonic changes, global climate transitions, and variations in orbital forcing (Maslin and Christensen 2007). The appearance of C4 pathway plants, signifying open savannah conditions represent global cooling and a long-term drying trend. Uplift related to formation of the East African Rift Valley resulted in changes in wind flow patterns from a more zonal to a more meridional direction. Evidence from lake, speleothem, and marine

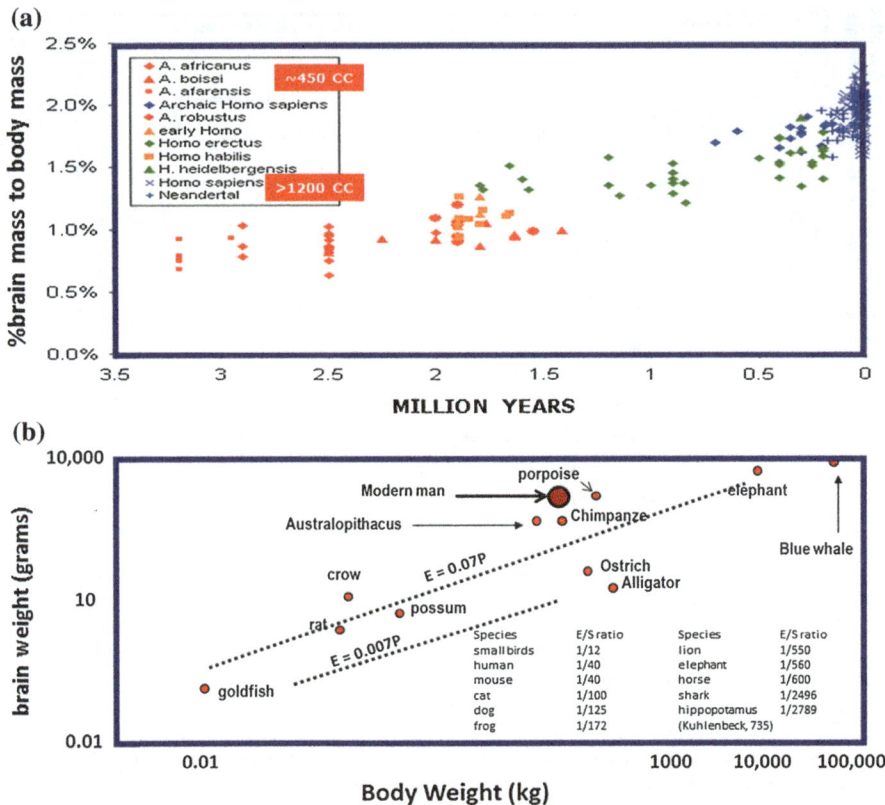

Fig. 6.3 a Evolution of the human brain. Brain mass as a percentage of body mass. (after Miguel and Henneberg (2001); Elsevier, by permission) processed by Nick Matzcke of NCSE http://www.pandasthumb.org/archives/2006/10/fun-with-homini-2.html http://www.pandasthumb.org/archives/images/Henneberg_de_Miguel_2004_Homo_hominins_single_lineage_fig1.png;
b Comparison between human brain weight/body mass weight and that of animals and birds. Reproduced with permission from http://www.brains.rad.msu.edu, and http://brainmuseum.org, supported by the US National Science Foundation. (Courtesy John Irwin Johnson. http://serendip.brynmawr.edu/bb/kinser/Int3.html www.neurophys.wisc.edu)

paleoclimate records manifest a long-term drying trend punctuated by episodes of short alternating periods of extreme wetness and aridity, involving appearance and disappearance of large, deep lakes (Figs 3.11, 3.12). The extreme climate variability has been related to the 41 kyr-long precession cycles and related wind patterns associated with the intensification of Northern Hemisphere glaciation (2.7–2.5 Ma), the Walker Circulation (1.9–1.7 Ma), mid-Pleistocene climate transition (1–0.7 Ma) and compression of the Intertropical Convergence Zone, resulting in rapid shifts from wet to dry conditions (Maslin and Christensen 2007; Trauth et al. 2007; Maslin and Trauth 2009). The extreme climate variability constituted catalyst for evolutionary change and has driven key speciation and

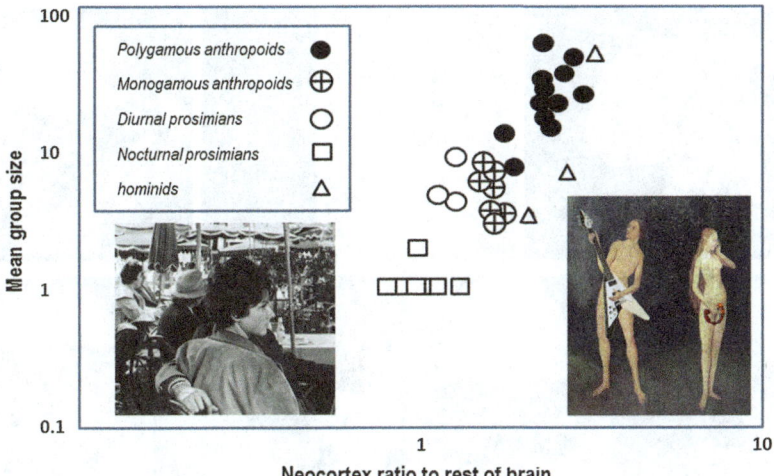

Fig. 6.4 Mean group size for individual genera plotted against neocortex ratio relative to rest of brain, including Polygamous anthropoids, monogamous anthropoids, diurnal prosimians (primates that include lemurs, lorises, bushbabies, and tarsiers), nocturnal prosimians and Hominins (Dunbar 1992, Fig. 1; Elsevier, by permission). Adam and Eve—Hieronymus Bosch http://www.flickr.com/photos/oddsock/92105811/ http://creativecommons.org/licenses/by/2.0/deed.en; Woman's profile—photograph by Arthur Glikson

dispersal events amongst mammals and hominins, which either originated or became extinct during climate upheavals.

Small human clans reacted to extreme climate changes during the Pleistocene—including cold fronts, storms, droughts and sea level changes—through migration within and out of Africa. Homo sapiens emerged during the glacial period which preceded the 124,000 years-old (124 kyr) Eemian interglacial, when temperatures rose temporarily by ~5 °C to nearly +1 °C higher than late-Holocene pre-industrial temperature, while sea levels were 6–8 m higher than the present (Hansen and Sato 2012). The Emergence of agriculture, and thereby of Neolithic human civilization, did not occur until the climate stabilized between ~10,000 and 7,000 years-ago, when large-scale irrigation along the great river valleys—the Nile, Euphrates, Indus and Yellow Rivers—became possible thanks to regulated river flows allowed by accretion and melting of snow in source mountain terrains.

It may never be known how fire was originally mastered, whether by percussion of flint stones or fast turning of wooden sticks, and whether this technique was developed in one or several places. During the earliest Paleolithic (~2.5 Ma) mean global temperatures were at least 2 °C warmer than the Holocene (Fig. 3.6b), which allowed human migration through open vegetated savannah in the Sahara and Arabian Peninsula. Confident evidence has emerged for the use of fire about one million years ago at the Wonderwerk Cave, Kuruman Hills, Cape Province, South Africa, from the study of palaeo-temperatures of burnt bones (Berna et al. 2012) (Fig. 6.6) (Table 6.1). Possible records of ~1.7–1.5 Ma-old fire places

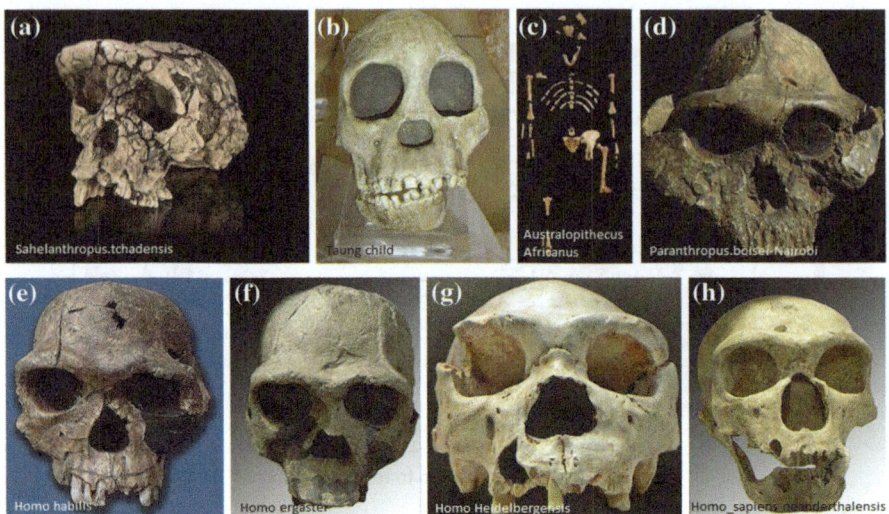

Fig. 6.5 Cranium of (**a**) Sahelanthropus.tchadensis. Cast of the Sahelanthropus tchadensis holotype cranium TM 266-01-060-1, dubbed Toumaï, in facio-lateral view. Author: Didier Descouens (http://en.wikipedia.org/wiki/File:Sahelanthropus_tchadensis_-_TM_266-01-060-1.jpg), **b** Taung Child. Australopithecus africanus male (replica) "Enfant de Taung" Discovered in Botswana 1924 by Raymond Dart (http://en.wikipedia.org/wiki/Taung_Child), **c** skeleton of Australopithecus Africanus (Lucy). Lucy skeleton (AL 288-1) Australopithecus afarensis, cast from Museum national d'histoire naturelle, Paris (http://en.wikipedia.org/wiki/File:Lucy_blackbg.jpg); **d** cranium of Paranthropus boisei-nairobi (http://en.wikipedia.org/wiki/File:Paranthropus boisei-Nairobi.JPG); **e** cranium of Homo habilis KNMR 1813 discovered at Koobi Fora author José-Manuel Benito Álvarez (España) Locutus Borg; (http://en.wikipedia.org/wiki/Homo_habilis); **f** cranium of Homo ergaster Skull KNM-ER 3733 discovered by Bernard Ngeneo in 1975 in Kenya (http://en.wikipedia.org/wiki/Homo_ergaster); **g** cranium of Homo heidelbergensis. Cranium 5 is one of the most important discoveries in the Sima de los Huesos, Atapuerca (Spain). The mandible of this cranium appeared, nearly intact, some years after its find, close to the same location. Author José-Manuel Benito (http://en.wikipedia.org/wiki/File:Homo_heidelbergensis-Cranium); **h** cranium of Homo sapiens-neanderthalis Skull discovered in 1908 at La Chapelle-aux-Saints (France). Author: Luna04 (http://en.wikipedia.org/wiki/File:Homo_sapiens_neanderthalensis.jpg) (Wikipedia/Creative Commons)

were recovered in excavations at Swartkrans (South Africa), Chesowanja (Kenya), Xihoudu (Shanxi Province, China) and Yuanmou (Yunnan Province, China) (Table 6.1). These included black, grey, and greyish-green discoloration of mammalian bones suggestive of burning. Clear evidence has been uncovered of the use of fire by Homo *erectus* and Homo *Heidelbergensis* at least 300,000 years-ago in Africa and the Middle East. Evidence for fire in sites as old as 1.4 Ma in Kenya and 750 kyr in France is controversial (Hovers and Kuhn 2004; Stevens 1989). Penetration of humans into central and northern Europe, including by *H. heidelbergensis* (600–400 kyr) and *H. neanderthalensis* (600–30 kyr) was

Fig. 6.6 Wonderwerk Cave: **a** Hand axes characteristic of the Acheulean of stratum; **b** Photograph of the east section in excavation where charred bones were found; **c** Representative micrograph of low-energy, water-bedded silt, sand, and 0.5-cm-thick gravel dolostone and flowstone where charred remains were found. (Berna et al. 2012, Figs. 1 and 2; Proceedings National Academy of Science USA, by permission)

facilitated by the use of fire for warmth, cooking and hunting. According to Roebroeks and Villa (2011) evidence for the use of fire, including rocks scarred by heat and burned bones, is absent in Europe until around 400 kyr, which implies humans penetrated northern latitudes even prior to the mastery of fire.

Fire allowed humans to migrate to harsh climate zones. Intensification of glacial-interglacial cycles drove intermittent dispersal of fauna, including humans, between Africa, the Middle East, southern and south-eastern Asia and southern Europe (Dennell and Roebroeks 2005) (Fig. 6.7), likely following the animals (Fig. 6.1). By at least ~1.8–1.6 Ma hominins arrived in Western Asia (Damanishi ~1.85 Ma), Eastern Asia (Yuanmou ~1.7 Ma; Nihewan ~1.66 Ma) and southeastern Asia (Sangiran (Java) ~1,66 Ma) (Smithsonian Institute 2012). Evidence for widespread use of fire in the late Paleolithic is indicated by charred logs, charcoal, reddened areas, carbonized grass stems and plants, and by wooden implements which may have been hardened by fire. Reliable evidence for the use of fire comes from the Bnot Yaakov Bridge, Israel, where between 790 and 690 kyr *H. erectus* or *H. ergaster* produced stone tools, butchered animals, gathered plant food and controlled fire (Goren-Inbar et al. 2004). Wood of six taxa was burned at the site, at least three of which are edible—olive, wild barley, and

Table 6.1 Prehistoric sites containing evidence or possible evidence of human use of fire

1.7 Ma	Yuanmou, Yunnan Province	Blackened mammal bones
1.5 Ma	Swartkan, South Africa	Burnt bones were found among Acheulean tools, bone tools, and bones with hominin-inflicted cut marks
1.42 Ma	Chesowanja, Kenya	Red clay shards. Heated to 400 °C to harden but re-interpreted as bushfire
1.5 Ma	Koobi Fora Kenya	Red clay shards heating at 200–400 °C
	Olorgesailie Kenya	"Hearth-like depression". Microscopic charcoal
	Gadeb, Ethiopia	Welded tuff near Acheulean artefacts
~1.0 Ma	Wonderwerk Cave, Northern Cape province, South Africa	Burned bone and ashed plant remain: carbonate-hydroxylapatite—undergoes characteristic recrystallization at approximately >500 °C
0.79–0.69 Ma	Bnot Yaakov bridge Israel	Burnt flintstones; *H. erectus* or *H. ergaster*
0.83–0.5 Ma	Java	*H. erectus* fossils; blackened bone and charcoal deposits
	Xihoudu in Shanxi Province	Evidence of burning by the black, grey, and greyish-green discoloration of mammalian bones
0.7–0.2 Ma	Cave of hearths in South Africa	
130000–120000 BP	Klasies river mouth	
110000–61000 BP	Kalambo falls in Zambia	Artefacts related to the use of fire by humans: charred logs, charcoal, reddened areas, carbonized grass stems and plants, and wooden implements which may have been hardened by fire
72000 BP	Stillbay culture. South Africa	Fire was used to heat treat silcrete stones to increase their workability before they were knapped into tools.
0.7–0.2 Ma	Cave of hearths in South Africa	

wild grape. The distribution of small burned flint fragments suggests that burning occurred in specific spots, possibly indicating hearth locations.

Wrangham (2009) attributed the increase in brain size and the drop in tooth size of *Homo erectus* (brain 900–1,200 cc) at 1.9–1.7 Ma, relative to *H. habilis* (brain 500–900 cc), to the onset of cooking of meat. Cooking allowed easier digestion of proteins, relieving early humans from energy-consuming chewing and enhancing the brain blood supply. It is from this stage that hominins grew taller and leaner shedding much of their original hair cover, allowing perspiration, cooling and the long range chase and hunt of animals.

Some of the best information regarding prehistoric fire cultures is related to burning strategies by native people in Africa (Laris 2002; Sheuyangea et al. 2005), North America (Stephens et al. 2007) and Australia. Aboriginal 'fires-tick

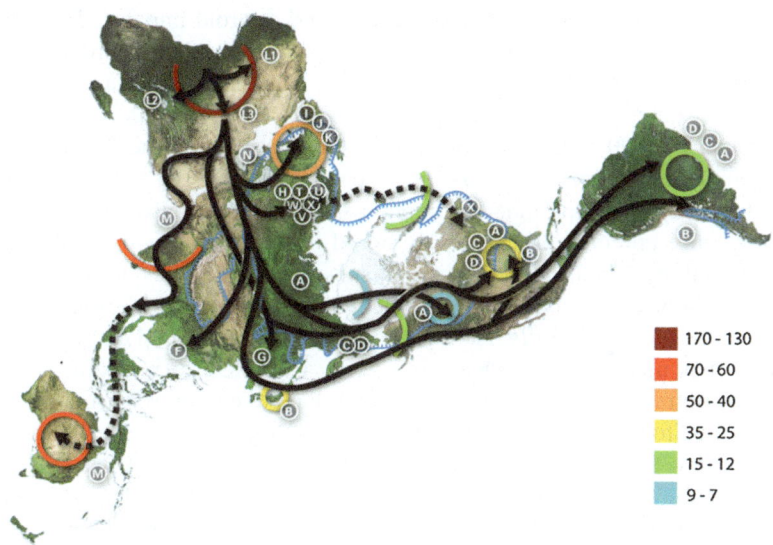

Fig. 6.7 A model of human migration based on mitochondrial DNA. The letters are the mitochondrial DNA haplogroups (pure motherly lineages); Haplogroups can be used to define fine genetic population often geographically oriented. The following are common divisions for mtDNA haplogroups: African: L, L1, L2, L3' Near Eastern: J, N; Southern European: J, K; General European: H, V; Northern European: T, U, X; Asian: A, B, C, D, E, F, G (note: M is composed of C, D, E, and G). Native American: A, B, C, D, and sometimes X. http://en.wikipedia.org/wiki/File:Map-of-human-migrations.jpg (http://www.mitomap.org/pub/MITOMAP/MitomapFigures/WorldMigrations.pdf) (license GNU Operating Systems: http://www.gnu.org/copyleft/fdl.html)

farming' associated with maintenance of small-scale habitat mosaics increased hunting productivity and foraging for small burrowing prey, including lizards (Gammage 2012). This led to extensive habitat changes, possibly including the extinction of mega-fauna, a debated issue (Miller 2005; Surovell and Grund 2012). The paucity of archaeological evidence for connections between hunting and mega-fauna extinction led some to invoke abrupt climate change, or even an extraterrestrial impact, for which there is no direct evidence, as drivers of the megafauna extinction. However, since the last glacial termination (LGT) bears close similarities with earlier terminations (Fig. 3.8), which mega-fauna survived, the LGT hardly explains the disappearance of mega-fauna such as diprotodons, giant kangaroos, marsupial tapirs, über-echidna (Surovell and Grund 2012; Johnson 2013).

Maori colonization of New Zealand 700–800 years-ago led to loss of half the South Island's temperate forest (McGlone and Wilmshurst 1999). These practices intensified upon European colonization, with extensive land cultivation and animal husbandry.

The harnessing of fire by humans, elevating the oxidizing capacity of the species by many orders of magnitude through release of solar energy stored by photosynthesis in plants (Kittel and Kroemer 1980), elevated planetary entropy to levels on a scale analogous to volcanic events and asteroid impacts. The effect on nature of anthropogenic combustion is tracking toward a similar order of magnitude. Thus, human respiration dissipates 2–10 calories per minute, a camp fire covering one square meter releases approximately $\sim 200{,}000$ Calories/minute, and the output of a 1,000 megawatt/hour power plant expends some 2.4 billion calories/minute, namely some 500×10^6 the mean energy level of human respiration.

The decrease in entropy inherent in photosynthetic reactions, which locally and transiently raise potential energy levels stored in plants, is dissipated through the energy output by plant oxidizing organisms, from microbes to herbivores to complex technological civilizations. Since the industrial revolution, further to decimating the world's bio-diverse forests, the oxidation of fossil carbon of ancient biospheres has increased the release of energy by orders of magnitude. In 2011 fossil fuel burning, cement production and land clearing released a total of 9.5 ± 0.5 GtC to the atmosphere (Global Carbon Project 2013), 54 % higher than in the 1990 Kyoto Protocol reference year. In 2011 coal burning was responsible for 43 % of the total emissions, oil 34 %, gas 18 %, and cement 5 %, with far reaching consequences for the atmosphere–ocean–cryosphere–biosphere system.

Inherent in ancient fire mythologies is the illegitimate nature of the acquisition of fire by humans, symbolized by Prometheus, the Titan who stole the fire from the gods breathing it into human clay figures (Frontispiece). According to the Rig Veda the hero Mātariśvan recovered fire which had been hidden from mankind. In Cherokee myth, after Possum and Buzzard had failed to steal fire, grandmother Spider used her web to sneak into the land of light and stole fire, hiding it in a clay pot. Among various Native American tribes of the Pacific Northwest and First Nations, fire was stolen and given to humans by Coyote, Beaver or Dog. According to some Yukon First Nations people, Crow stole fire from a volcano in the middle of the water. According to the Creek Indians, Rabbit stole fire from the Weasels. In Algonquin myth, Rabbit stole fire from an old man and his two daughters. In Ojibwa myth, Nanabozho the hare stole fire and gave it to humans. In Polynesian myth, Maui stole fire from the Mudhens. In the Book of Enoch, the fallen angels and Azazel teach early mankind to use tools and fire. Because fire allows humans warmth, protection from animals, cooking, pottery and migration to cold parts of the planet, the underlying warnings implicit in these legends is perplexing. Over the millennia, culminating in the Anthropocene (Ruddiman 2003; Steffen et al. 2007), however, the hidden meaning of these forewarnings has become progressively manifest.

The effects of fire on the human mind are known to those who have camped for long periods around campfires, as this author has on-and-off for 40 years. For hundreds of thousands of years, gathered during long nights around camp fires, captivated by the flickering life-like dance of the flames, humans developed imagination, insights, cravings, fears, premonitions of death and, thereby, aspirations of immortality, omniscience, omnipotence and concepts of supernatural being, or gods. Pantheistic beliefs revered the Earth, its rocks and living creatures.

Fear, an instinctive sense arising in animals when endangered, is created in the human mind allowing it to foresee risks in advance (Fig. 6.8). Since at least

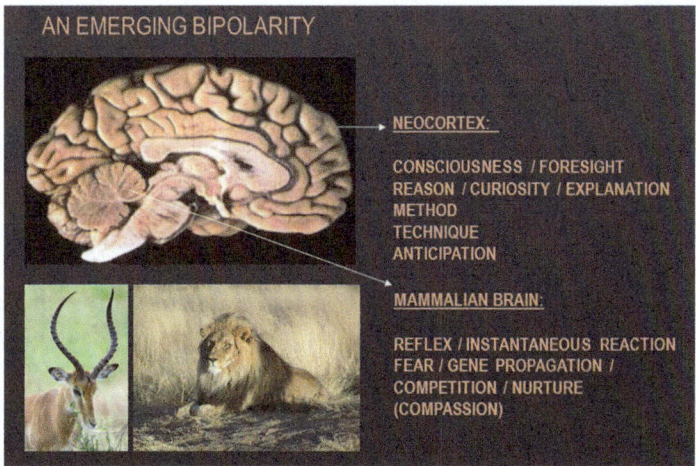

Fig. 6.8 An emerging bipolarity—the conflict between the Neocortex (consciousness, foresight and reason) and the mammalian brain (reflex/instantaneous reaction). On the surface of it no such conflict exists among animals, operating on instinct when faced with danger. Impala—Young male Impala Mikumi National Park, Tanzania Photographer: Muhammad Mahdi Karim http://en.wikipedia.org/wiki/File:Male_impala_profile.jpg;Lion (Panthera leo) lying down in Namibia. Photographer: Kevin Pluck (http://commons.wikimedia.org/wiki/File:Lion_waiting_in_Namibia.jpg)

Fig. 6.9 Mungo man No. 3, the skeleton of a *tall* middle-aged man, was excavated in March, 1974, by a team led by the late Alan Thorne (Courtesy Colin Groves)

130,000 years-ago the recognition of death and a yearning for immortality have been expressed in ritual burial (Fig. 6.9). An example is the Skhul cave, Mount Carmel, Israel, where skeletons painted in red ochre are surrounded by tools (Hovers and Kuhn 2004). Cremation constituted a special way of merging the spirit of the deceased with fire, allowing a passage of the soul to eternity. When civilization rose, burial rituals evolved into grand monuments for the afterlife, represented by the Egyptian pyramids (Fig. 7.7) and Chinese imperial burial caves, the latter including entombed entourages intended to serve the ruler in the hereafter, such as Emperor Qin Shi Huang's terracotta army (Wood 2008).

The human premonition of death has led to tension between foresights acquired by the neocortex and the instant reflexes of the mammalian brain (Koestler 1986) (Fig. 6.8). Where the cerebral neocortex invents tools and techniques and identifies future dangers, the primitive brain reacts instinctively through defensive/aggressive impulses (Dawkins 1976), with ensuing conflict. When equipped with weapons designed by the intelligent brain, such responses lead to destructive violence, translated from the individual scale into tribal and global wars. By analogy to infanticide experienced by rival baboons, running through human history are mass sacrifices of the young, children thrown into the Moloch's fire, the Maya blood cult (Clendenin 1995), all the way to generational sacrifice as in World War I and II—called '*war*', under tribal or religious flags, appeasing the angry gods in a quest for immortality.

A Maya war song (1300–1521 AD):

There is nothing like death in war,
Nothing like flowering death, so precious to him who gives life,
Far off I see it, my heart yearns for it.

With the onset of space exploration, from the Sputnik to the lunar landings, the Galileo space craft and Voyager's interstellar mission, religious mythologies evolved into a space cult, alluding to colonization of the planets where, presumably, *H. sapiens* would proceed to overwhelm new environments. The suggestion by space race proponents as if life has been seeded on Earth by comets appears to be as ideologically motivated as it is lacking in evidence. While hosting amino acids, comets and space dust are not known to contain biomolecules. The allusion of the space cult to panspermia and human propagation to other planets and solar systems ignores the central observation arising from planetary exploration—barring possible presence presence of microbes—good planets are hard to come by.

Ancient DNA sequences suggest that earlier than 400 kyr-ago Homo species diverged from the ancestors of modern humans (Endicott et al. 2010). *Homo sapiens* is thought to have evolved in Africa during the glacial period which preceded the Eemian interglacial period ($\sim$130–114 kyr) (Klein 2009). During this period and early parts of the last glacial anatomically near-modern humans occupied the Levant (Bar Yosef 2000). However, it is only after 50 kyr that humans of near-modern behavior occur in the archaeological record in the Near East. The major expansion of *H. sapiens* out-of-Africa is considered to have occurred between 60 and 45 kyr (Henn et al. 2012). Human expansions outside of Africa were likely associated with climatic fluctuations (Stewart and Stringer 2012). Significantly this migration occurred following the coldest temperatures recorded in Greenland and Antarctic ice cores and related to the sun-blocking effects of the Toba volcanic eruption, dated by $^{40}Ar/^{39}Ar$ as 73.88 ± 0.32 kyr

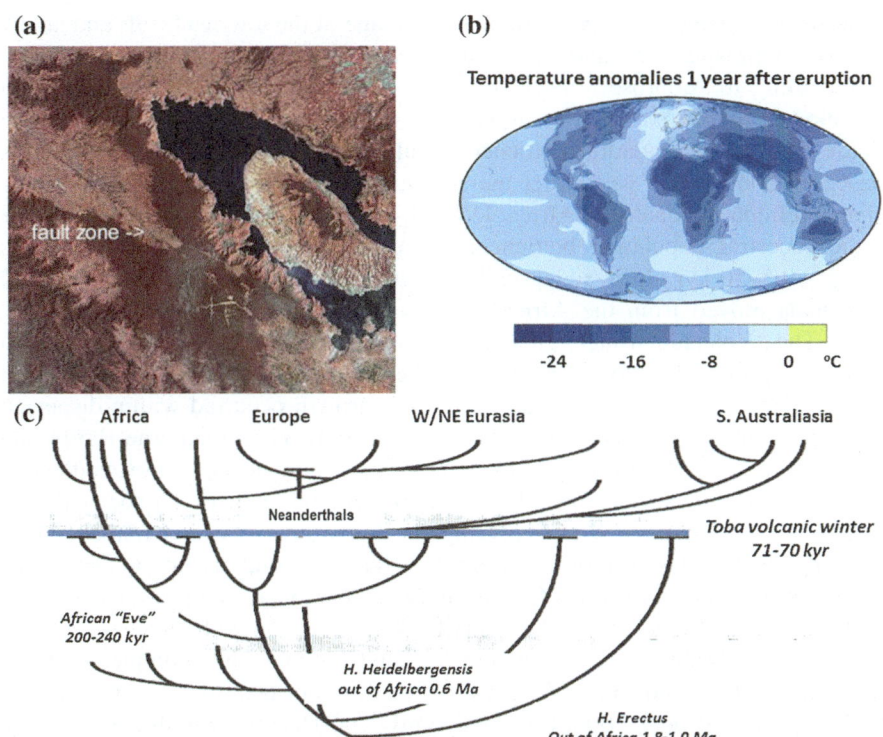

Fig. 6.10 The Toba eruption of ~77–69 kyr (thousands years) ago: **a** Landsat image of Toba caldera (Landsat Pathfinder Project); **b** Annual near-surface temperature anomalies for the year following a super-volcanic eruption like the Toba eruption if it were to occur today. Most land areas cool by 12 °C compared to average. Some areas, like Africa, cool by 16 °C (Jones et al. 2005; Springer, by permission); **c** The Volcanic Winter/Weak Garden of Eden model proposed by Ambrose (1998). Population subdivision due to dispersal within African and to other continents during the early Late Pleistocene is followed by bottlenecks caused by volcanic winter, resulting from the eruption of Toba about 71 kyr. The bottleneck may have lasted either 1,000 years, during the hyper-cold stadial period between Dansgaard-Oeschger events 19 and 20, or for 10 kyr during oxygen isotope stage 4. Population bottlenecks and releases are both synchronous. More individuals survived in Africa because tropical refugia were largest there, resulting in greater genetic diversity in Africa (Ambrose 1998, Fig. 4; Elsevier, by permission)

(Storey et al. 2012) (Fig. 6.10). The eruption is followed by ~10 °C drop in Greenland surface temperature over ~150 years, related to the climatic impact of volcanic aerosols. The Toba super-eruption is estimated to have produced some 2,500–3,000 km^3 of lava and pyroclastics, probably injecting at least 10^{15} gr of fine ash into the stratosphere (Rampino and Self 1993; Gathorne-Hardy and Harcourt-Smith 2003). Ash and sulfate aerosols were deposited in both hemispheres, forming a time-marker horizon that can be used to synchronize late Quaternary records globally (Storey et al. 2012). The discovery of stone artifacts covered by Toba volcanic ash in Malaysia and India indicates that by 74 kyr *H. sapiens* or Denisovans have reached Southeast Asia (Petraglia et al. 2007; Timmreck et al. 2012). If so, survivors of the Toba eruption may have been able to migrate eastwards following the eruption, taking advantage of the low sea levels and newly exposed continental shelf and land bridges (Storey et al. 2012) (Fig. 6.7).

According to Ambrose (1998) Toba's volcanic winter could have decimated most modern human populations, especially outside isolated tropical enclaves. The release from this population bottleneck could have occurred at the end of this phase, or about 10,000 years after the eruption, with most survivors found in the tropical refugia in equatorial Africa. DNA re-sequencing studies estimate the size of the ancestral population bottleneck in Africa as 12,800–14,400 individuals during ~65–50 kyr. It is suggested that small groups of only ~1,000–2,500 individuals moved from the African continent into the Near East (Henn et al. 2012). The 'Volcanic winter' effects of Toba may have reduced populations to levels low enough for 'founder effects', where local adaptations produced rapid differentiation at about 70 kyr. Around 50 kyr growth occurred within dispersed populations that were genetically isolated from each other. Gathorne-Hardy and Harcourt-Smith (2003) raise questions regarding a human population bottleneck following Toba and suggest there has been no mammal extinction associated with Toba. Human genome studies analyzing the Y-chromosome and mitochondrial DNA support the existence of a bottleneck in human evolution but the age of this bottleneck remains uncertain. Consistent with this view, archaeological finds (McBrearty and Brooks 2000) show that by 73.5 Kyr modern humans occupied a diverse range of habitats over the whole African Continent, with complex tool-kits and ability to hunt and forage for a variety of different taxa over extensive areas (McBrearty and Brooks 2000; Gathorne-Hardy and Harcourt-Smith 2003).

By about 20,000 years-ago, toward the end of the last glacial age, *Homo sapiens* spread over much of Asia, Australia and Europe. From paleo-anthropological and DNA evidence, the species interbred and replaced archaic human species, including the Neanderthals in Europe, the Denisovans in Siberia and Homo floresiensis on the Indonesian island of Flores (Stewart and Stringer 2012). These authors pointed to the relation between the phylo-geographic (genetic bio-geographical), paleontological and DNA records of extant animals and human migration patterns by the end-Pleistocene, including evidence of extinctions. The genetics of human parasites, morphology and linguistics suggest the *H. sapiens* takeover involved an overall loss of genetic diversity. Thus, whereas genomes from African populations retain a large number of unique variants, a dramatic reduction is observed in genetic diversity of people outside of Africa (Henn et al. 2012).

References

Ambrose SH (1998) Late Pleistocene human bottlenecks, volcanic winter, and differentiation of modern humans. J Hum Evol 34:623–651
Bar-Yosef O (2000) The Middle and early upper Paleolithic in Southwest Asia and neighboring regions In: Bar-Yosef O, Pilbeam D (eds) The geography of Neanderthals and Modern Humans in Europe and the Greater Mediterranean. Peabody Museum Press, Cambridge, p 107–156

Berna F, Goldberg P, Horwitz LK, Brink J, Holt S, Bamford M, Chazang M (2012) Microstratigraphic evidence of in situ fire in the Acheulean strata of Wonderwerk Cave, Northern Cape province. South Africa. Proc Nat Acad Sci 109(20):E1215–E1220

Clendinnen I (1995) Aztecs: an interpretation. Cambridge University Press, Cambridge, p 389

Dawkins R (1976) The selfish gene. Oxford University Press, Oxford, p 384

deMenocal PB (2004) African climate change and faunal evolution during the Pliocene-Pleistocene. Earth Planet Sci Lett 220:3–24

Dennell RW, Roebroeks W (2005) Out of Africa: An Asian perspective on early human dispersal from Africa. Nature 438:1099–1104

Dunbar RIM (1992) Neocortex size as a constraint on group size in primates. J Hum Evol 22:469–493

Dunbar RIM (1996) Grooming, gossip and the evolution of language. Faber and Faber, London, p 219

Endicott P, Ho SYW, Stringer C (2010) Using genetic evidence to evaluate four palaeoanthropological hypotheses for the timing of Neanderthal and modern human origins. J Hum Evol 59:87–95

Gammage B (2012) The biggest Estate on Earth: how Aborigines made Australia. Allen and Unwin, Crows Nest, p 384

Gathorne-Hardy FJ, Harcourt-Smith WEH (2003) The super-eruption of Toba, did it cause a human bottleneck? J Hum Evol 45:227–230

Global Carbon Project, Global Carbon Budget (2013) http://www.globalcarbonproject.org/carbonbudget/12/hl-full.htm

Goren-Inbar N, Alperson N, Kislev ME, Simchoni O, Melamed Y, Ben-Nun A, Werker E (2004) Evidence of Hominin Control of Fire at Gesher Benot Ya'aqov, Israel. Science 304:725–727

Griffin DR (1992) Animal minds. University of Chicago Press, Chicago

Groves C (1993) Our earliest ancestors. In: Burenhult G (ed) The first humans: human origins and history to 10,000 BC. Harper-Collins Publishers, New York, p 33–40, 42–45, 47–52

Hansen JE, Sato M (2012) Paleoclimate implications for human-made climate change. Clim Change 2012:21–47

Henn BM, Cavalli-Sforza LL, Feldman MW (2012) The great human expansion. Proc Nat Acad Sci 109:17758–17764

Hovers E, Kuhn S (2004) Transitions before the transition: evolution and stability in the middle Palaeolithic and Middle Stone Age. Springer, New York, p 171–188

Johnson C (2013) Hunting or climate change? Megafauna extinction debate narrows. http://theconversation.com/hunting-or-climate-change-megafauna-extinction-debate-narrows-10602#comment_156216

Jones GS, Gregory JM, Stott PA, Tett SFB, Thorpe RB (2005) An AOGCM simulation of the climate response to a volcanic super-eruption. Clim Dyn 25:725–738

Kittel C, Kroemer H (1980) Thermal physics, 2nd edn. W.H. Freeman and Company, San Francisco

Klein R (2009) The human career: human biological and cultural origins. University Chicago Press, Chicago

Klein RG, Edgar B (2002) The dawn of human culture. Wiley, New York, p 288

Koestler A (1986) Janus: a summing up. Picador Books, UK

Laris P (2002) Burning the seasonal mosaic: preventative burning strategies in the wooded savanna of Southern Mali. Human Ecol 30:155–186

Maslin MA, Christensen B (2007) Tectonics, orbital forcing, global climate change, and human evolution in Africa: introduction to the African paleoclimate special volume. J Hum Evol 53(5):443–464

Maslin MA, Trauth MH (2009) Plio-Pleistocene East African pulsed climate variability and its influence on early human evolution. In: The first humans—Origin and early evolution of the Genus Homo. In: Grine GE, Fleagle JG, Leakey RE (eds) Verteb paleobiology paleoanthropology, Springer, Netherlands, p 151–158

McBrearty S, Brooks AS (2000) The revolution that wasn't: a new interpretation of the origin of modern human behavior. J Hum Evol 39:453–563
McGlone MS, Wilmshurst J (1999) Dating initial Maori environmental impacts in New Zealand. J Quarter Sci 59:5–16
McHenry HM (2009) Human evolution. In: Ruse M, Travis J (eds) Evolution: the first four billion years. The Belknap Press of Harvard University Press, Cambridge, p 265
Miguel CD, Henneberg M (2001) Variation in hominid brain size: how much is due to method? HOMO-J Comp Hum Biol 52:3–58
Miller GH (2005) Ecosystem collapse in Pleistocene Australia and a human role in megafaunal extinction. Science 309:287–290
Narby J (2005) Intelligence in nature. Penguin, New York
Petraglia M et al (2007) Middle Paleolithic assemblages from the Indian subcontinent before and after the Toba super-eruption. Science 317:114–116
Rampino MR, Self S (1993) Climate-volcanism feedback and the Toba eruption of ~74,000 years-ago. Quat Res 40:269–280
Roebroeks W, Villa P (2011) On the earliest evidence for habitual use of fire in Europe. http://johnhawks.net/node/15346
Ruddiman WF (2003) Orbital insolation, ice volume, and greenhouse gases. Quatern Sci Rev 22:1597–1629
Sheuyangea A, Oba G, Weladji RB (2005) Effects of anthropogenic fire history on savanna vegetation in northeastern Namibia. J Environ Manag 75:189–198
Smithsonian Institute (2012) What does it mean to be human? http://humanorigins.si.edu/research/asian-research/hobbits; http://humanorigins.si.edu/research/asian-research/earliest-humans-china
Steffen W, Crutzen PJ, McNeill JR (2007) The Anthropocene: are humans now overwhelming the great forces of nature? Ambio 36:614–621
Stephens SL, Martin RE, Clinton NE (2007) Prehistoric fire area and emissions from California's forests, woodlands, shrub lands and grasslands. Forest Ecol Manage 251:205–216
Stevens JR (1989) Hominid use of fire in the lower and middle Pleistocene: a review of the evidence. Current Anthrop Univ Chicago Press 30:1–26
Stewart JR, Stringer CB (2012) Human evolution out of Africa: the role of Refugia. Climate Change Sci 335:1317–1321
Storey M, Roberts RG, Saidin M (2012) Astronomically calibrated 40Ar/39Ar age for the Toba super-eruption and global synchronization of late Quaternary records. Proc Natl Acad Sci 109:18684–18688
Surovell TA, Grund BS (2012) The Associational Critique of Quaternary Overkill and Why It is Largely Irrelevant to the Extinction Debate. Am Antiq 77(4):672–687
Timmreck C et al (2012) Climate response to the Toba super-eruption: regional changes. Quatern Int 258:30–44
Trauth MH, Maslin MA, Deino ALStrecker MRBergner AGN, Duhnforth M (2007) High- and low-latitude forcing of Plio-Pleistocene East African climate and human evolution. J Hum Evol 53:475–486
Wood F (2008) China's first Emperor and his Terracotta Warriors. Macmillan publishing, London
Wrangham R (2009) Catching fire: how cooking made us human. Basic Books, New York, p 320
Zimmer C (2005) Smithsonian intimate guide to human origins. Madison Press Books, Totonto, p 176

Chapter 7
Climate and Holocene Civilizations

Abstract Since the Neolithic and throughout history cultivation and agriculture-based civilizations concentrated along rivers, or above groundwater reservoirs, depended critically on availability of water, which in turn depended on the climate, including annual river rhythms, the effects of forests on microclimate, soil erosion, and in some parts of the world such as southeast Asia on volcanic regimes.

In the wake of the last deglaciation (Fig. 7.1) peak pre-Holocene temperatures are represented by the ~14.7–12.9 kyr Bolling-Alerod interstadial, preceded by a cool *'Older dryas'* phase and followed by sharp cooling at the *'Younger dryas'* at 12.9–11.7 kyr (Fig. 7.2), succeeded by the Holocene Optimum (Shakun et al. 2012; Steffensen et al. 2008). Peak early Holocene conditions involved heavy precipitation and thereby rates of erosion (Clift et al. 2007), perhaps echoed by Noah's Ark story. At ~8.2 kyr a sharp stadial involving temperature decline of several degrees Celsius in the North Atlantic was associated with discharge of cold water from the Laurentian ice sheet through Lake Agassiz (Wagner et al. 2002; Lewis et al. 2012; Wiersma et al. 2011) (Fig. 7.3). High temperatures during the Holocene maximum resulted in strong East African monsoons and higher rainfall in the Sahara relative to previous and succeeding periods, which explains the presence of animals like giraffes and gazelle recorded in ancient rock paintings (Fig. 7.4). Consequently cooler and drier conditions ensued with temperature decline in the range of −1 to −5 °C in mid-latitudes and of −3 °C from ancient coral reef in Indonesia. Atmospheric CO_2 declined by ~25 ppm over ~300 years. About 8.2 kyr the earliest settled human communities in Catal Huyuk (southern Anatolia) were abandoned, likely due to droughts associated with the cooling, and were not reoccupied until about five centuries later when climate improved.

A decline in CO_2 and methane following the Holocene Optimum at ~8–6 kyr was followed by a slow rise in CO_2 from ~7,000 BP and methane from ~5,000 BP (Fig. 7.5). According to Ruddiman (2003) the natural interglacial cycle has been overprinted by Neolithic burning and land clearing, halting a decline in CO_2 and methane and thereby an onset of the next glacial (Kutzbach et al. 2010) (Fig. 7.6). Other authors regard the mid-Holocene rise of greenhouse gases as a natural perturbation in the interglacial, comparable with features of the 420–405 kyr

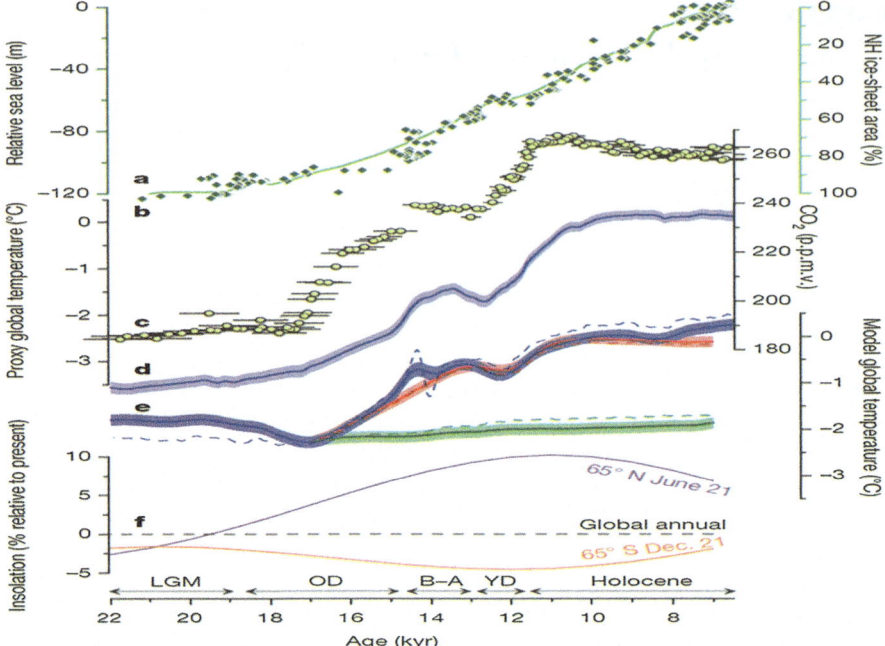

Fig. 7.1 Global temperature and climate forcings: **a** Relative sea level (*diamonds*); **b** Northern Hemisphere ice-sheet area (*line*) derived from summing the extents of the Laurentide, Cordilleran and Scandinavian ice sheets through time; **c** Atmospheric CO_2 concentration **d** Global proxy temperature stack; **e** Modeled global temperature stacks from simulations. *Dashed lines* show global mean temperatures in the simulations, using sea surface temperatures over ocean and surface air temperatures over land; **f** Insolation forcing for latitudes 65°N (*purple*) and 65°S (*orange*) at the local summer solstice, and global mean annual insolation (*dashed black*) (Shakun et al. 2012, Fig. 3; Nature, by permission)

Holsteinian interglacial (Broecker and Stocker 2006). The late Holocene is interrupted by mild warming phase ~900–1400 AD years-ago (Medieval Warm Period) and a cool phase during ~1550–1800 AD (Little Ice Age). Broadly these correspond to periods of solar insolation reaching + 0.5 watt/m^2 and −0.5 watt/m^2 relative to pre-industrial levels, respectively, attributed in part to changes in insolation related to sunspot activity (Solanki 2002).

In the wake of climate disruptions associated with the last glacial termination about 14–10 kyr ago, the climate stabilized between ~7 and 5 kyr when sea level reached maximum (Fig. 7.1). Of all the factors which underlie the rise and fall of ancient civilizations along the Nile, Tigris, Euphrates, Indus and the Yellow River valleys, the main control was exerted by the seasonally regulated balance in their mountain source regions between accumulation and melting of snow. Thus, cold spells would decrease river flow, ensuing in droughts, whereas strong

7 Climate and Holocene Civilizations

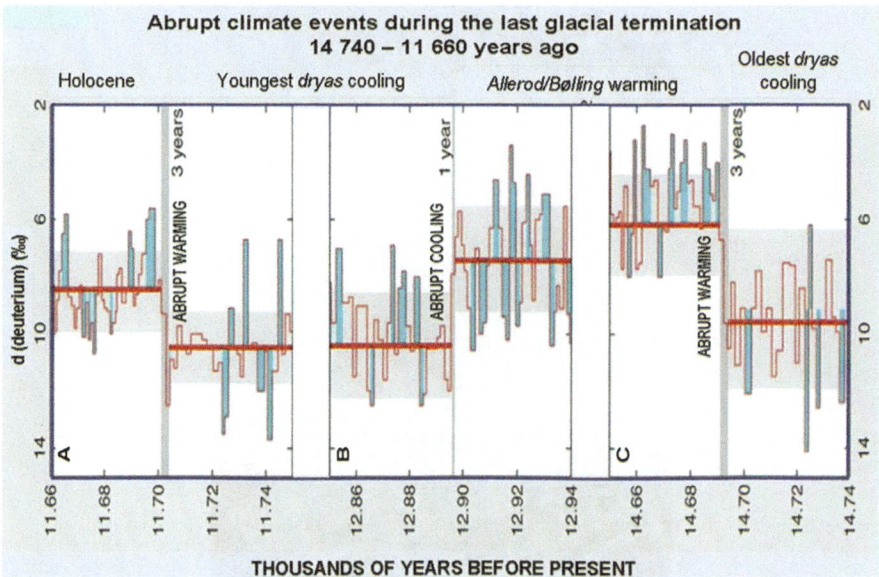

Fig. 7.2 d (deuterium) ‰ variations during the last glacial termination 14,740–11,660 years-ago, marking the oldest dryas cooling, Allerod/Bolling warming, Youngest dryas cooling and Holocene optimum. The mean values are shown as bold red lines. Note the abrupt shifts over periods of 1–3 years between climate states (Steffensen et al. 2008, Fig. 3; American Association for Advancement of Science, by permission)

monsoons result in floods and erosion of terraces. Under stable rhythmic climate, seasonal regulation of river flow accompanied with deposition of fertile silt allowed river terrace cultivation, providing food for villages, towns and subsequently kingdoms and empires. The Nile River, fed by water from the Ethiopian Mountains, allowed a flourishing of the Old Kingdom (4660–4160 BP), Middle Kingdom (4040–3640 BP) and the New Kingdom (3550–3070 BP) (Fig. 7.7). The largest pyramids were built during the Old Kingdom. The greatest expansion of the Pharaoh's territories in the Middle East occurred during the New Kingdom.

Stages in the history of the Nile River included:

- ~20–12.5 kyr—Northeast Africa—frozen Ethiopia Mountains; stable sediment alluviation and terrace building by a low-flow river; Hunter-gatherers.
- ~12.5–8 kyr—Northeast Africa—high floods (the 'wild Nile') due to heavy rains in the Ethiopia Highlands; little rain along the Nile. Increased vegetation at the source leads to less sedimentation and thus greater erosion of river terraces; Near-disappearance of population.
- ~8–6 kyr—Seasonal climate, stabilization of the Nile and re-aggradation of alluvial terraces, allowing irrigated agriculture.

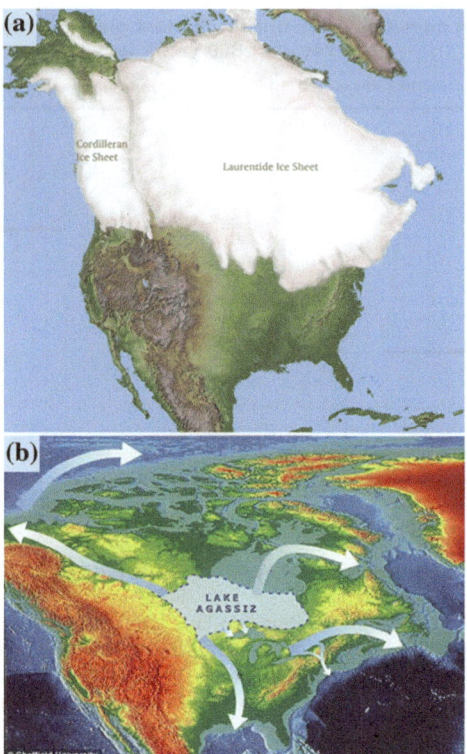

Fig. 7.3 a Original extent of the Laurentian ice sheet (Courtesy: Randall Carlson—http://www.cosmographicresearch.org/prelim_glacial_maximum.htm); **b** Reconstruction of Lake Agassiz outflows south of the Laurentian ice sheet (Sheffield University based on NOAA ETOPO2 data; http://www.sheffield.ac.uk/geography/about/100401, by permission)

- ~7.5–5.1 kyr pre-dynastic Egypt.
- ~5.5 Kyr—Retreat of the rain belt southward.
- ~4.686–4.181 kyr—Old Kingdom.
- ~4.2–4.0 kyr—Desertification.
- ~4.0–3.7 kyr—Middle Kingdom.
- ~3.570–3.069 kyr—New Kingdom.
- ~3.2–2.55 kyr—Iron age cold period.

The 4.1 kyr desertification constituted one of the most severe climatic events of the Holocene period in terms of impact on civilization. It is very likely to have caused the collapse of the Old Kingdom in Egypt as well as the Acadian Empire in Mesopotamia, and Indus Harappan cultural domain. Radiocarbon age determination from Tell Leilan, northeast Syria, uncovers evidence for an incipient collapse of the Acadian empire near 4170 ± 150 BP (Weiss et al. 1993). Deep sea core sediments from the Gulf of Oman testify to a several-fold increase in wind-borne

Fig. 7.4 Sahara peak Holocene rock painting. Tadrart Acacus form a mountain range in the Sahara desert of western Libya. The area is known for its rock paintings dating from 12,000 BC to 100 AD. The paintings reflect the changing environment of the Sahara desert which used to have a much wetter climate. Nine thousand years-ago the surroundings were green with lakes and forests and with large herds of wild animals as demonstrated by rock paintings at Tadrart Aracus of animals such as giraffes, elephants and ostriches. Photographer: Roberto D'Angelo (roberdan); Wikipedia Commons http://en.wikipedia.org/wiki/File:Tadrart_Acacus_1.jpg

aeolian components from 4025 ± 125 BP, representing development of arid conditions in the source regions of the dust in Mesopotamia (Cullen et al. 2000).

The Tigris and Euphrates rivers, fed by the waters from the Taurus Mountains, constitute the cradle of the Mesopotamian ("Land between the rivers") civilization, where irrigation developed from about 6000 BP and Sumer cities grew between 3200 and 2350 BP, succeeded by Babylon. The Harappan civilization was developed by Dravidians people along the Indus River, fed from the Himalaya. Cultivation along the Yellow and Yangzi Rivers including the Xia, Shang and the Zhou Dynasties developed from about 7000 BP.

According to deMenocal (2001) late Holocene climate perturbations included repeated inter-annual droughts and infrequent decadal droughts. Multi-decade-long to multi-century-long droughts were rare but formed integral components of the natural climate variability. Paleoclimate and archaeological records demonstrate close relations between prolonged droughts and social collapse. Overpopulation, deforestation, resource depletion and warfare reinforced social collapse. Repeated droughts likely constituted the root factors in the downfall of the Acadian, May, Mochica and Tiwanaku civilizations.

A dry period in central America during 530–650 AD, followed by droughts during 800–1000 AD, recorded by high gypsum precipitation and high $\delta^{18}O$ in lake sediments (Lakes Chichancanab and Punta Laguna) led to collapse of the Maya civilization between 750 and 790 AD (Fig. 7.8). The last Maya monument

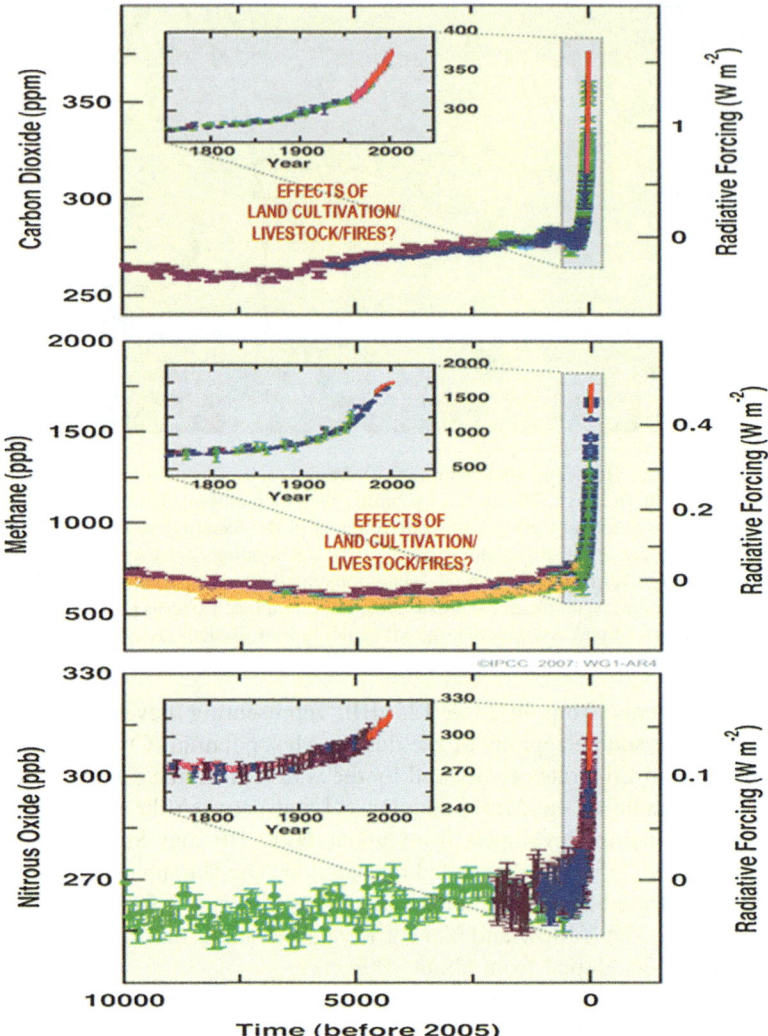

Fig. 7.5 The concentrations and radiative forcing by (**a**) carbon dioxide (CO_2) (**b**) methane (CH_4) and (**c**) nitrous oxide (N_2O) over the last 10,000 years reconstructed from Antarctic and Greenland ice and firn data (symbols) and direct atmospheric measurements (panels a,b,c, *red lines*). The Physical Science Basis. Working Group I Contribution to the Fourth Assessment Report of the Intergovernmental Panel on Climate Change. Cambridge University Press. Figure TS-2, by permission) http://www.ipcc.ch/publications_and_data/ar4/wg1/en/figure-ts-2.html)

was constructed in 990 AD (deMenocal 2001). Further confirmation of these trends is based on a paleoclimate study of Balum Cave deposits, Belize, correlated with dated Maya stone monuments (Shen 2012). The study indicates a high rainfall period during 440–660 AD followed by a drying trend between 660 and 1000 AD, leading to collapse between 1020 and 1100 AD.

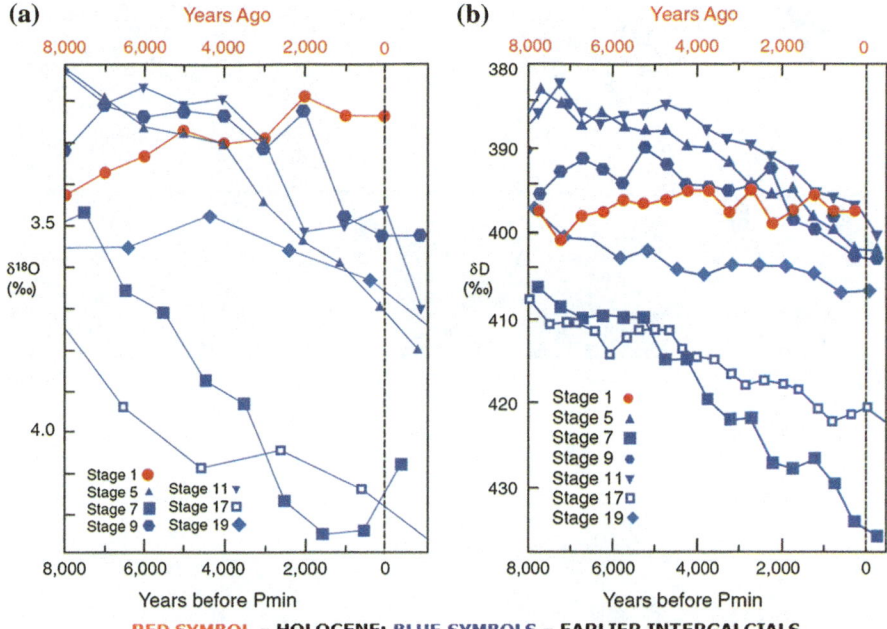

Fig. 7.6 The early Anthropocene hypothesis: **a** Isotopic $\delta^{18}O$ evidence for an anomalous warming trend in the late Holocene based on studies of benthic foraminifera (Lisiecki and Raymo 2005) showing a trend toward lighter $\delta^{18}O$ (warmer) values in the late Holocene, but trends toward heavier $\delta^{18}O$ (colder) values during earlier inter-glaciations; **b** Deuterium (δD) trend at Dome C Antarctica (Jouzel et al. 2007) shows a stable warming trend during the late Holocene, but trends toward lighter (colder) values during earlier inter-glaciations (Kutzbach et al. 2010, Fig. 4; Jouzel et al., 2007; Springer, by permission; American Association for the Advancement of Science, by permission)

A decrease in temperatures during 600–1000 AD is also recorded in the Quelccaya ice cores in Peru, showing an increase in the accumulation of ice and dust particle, signifying colder climate and a decrease in precipitation which affected the Tiwanaku civilization (300–1100 AD), leading to collapse between 1100 and 1400 AD. Based on the study of tree rings, during the 14–15th centuries the Khmer Empire in Cambodia experienced decades-long droughts induced by strong El-Nino events, interspersed with intense monsoons. Increased climate variability, which damaged water supply dams and canals on which the Angkor Watt civilization depended, led to the demise of Khmer civilization (Fig. 7.9) (Buckley et al. 2010).

Fig. 7.7 The rise of ancient river civilizations—Egypt and Babylon. **a** The wheat harvest Deir el-Medina, tomb of Sennutem; author: Egyptian tomb artist (http://upload.wikimedia.org/wikipedia/commons/a/ae/Egyptian_harvest.jpg); **b** The Giza Cheops Pyramids 4589–4566 BP, 2,300,000 giant stone blocks over 2.5 tons, built by 100,000 slaves. http://en.wikipedia.org/wiki/File:All_Gizah_Pyramids.jpg; **c** Tower of Babel, by Gustav Doré, 1866. Wikipedia Commons http://en.wikipedia.org/wiki/File:Confusion_of_Tongues.png

The end of the 1st millennium and the first half of the 2nd millennium constituted the *Medieval Warm Period* (MWP—900–1200 AD), which was about +0.4 °C warmer than the period ∼1500–1900 AD (Fig. 7.10). The coolest century was the 17th (−0.4 °C relative to pre-industrial temperatures), including the *Little Ice Age* ∼1600–1700 AD related to a decline to near-absence of sun spots (Jones et al. 2001; Solanki 2002). Multi-proxy temperature reconstructions for the northern hemisphere show that the recent 30-year period is likely to have been the warmest of the millennium (Jones et al. 2001). Southern Hemisphere temperature reconstructions indicate cooler conditions before 1900 but the data are less reliable

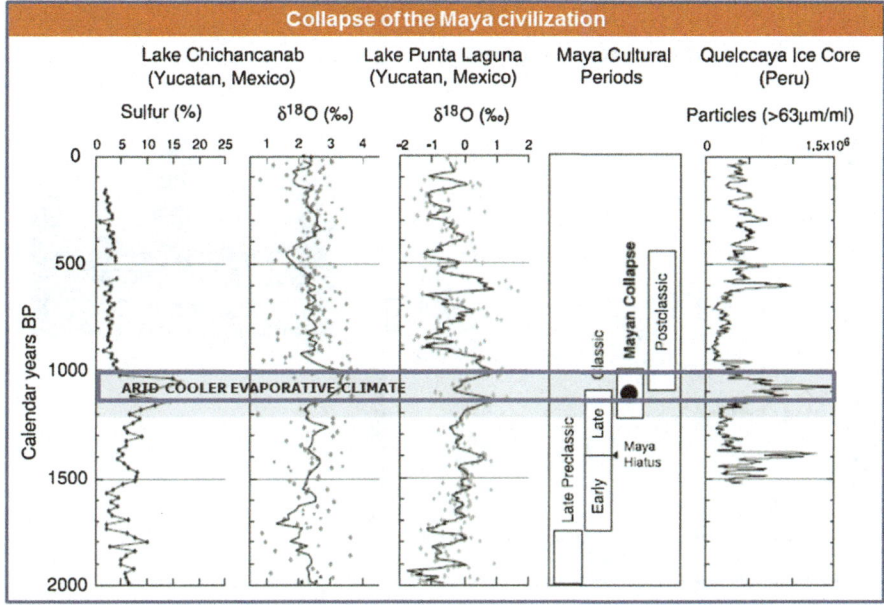

Fig. 7.8 Incipient collapse of the Classic Maya civilization near 750–790 A.D. The last Maya monument construction has been dated at 909 A.D. from the Maya Long Count inscriptions. Well-dated sediment cores from Lakes Chichancanab and Punta Laguna (northern Yucatan Peninsula, Mexico) document an abrupt onset of more arid conditions spanning 200 years between 800 and 1000 A.D., as evidenced by more evaporative (higher) $\delta^{18}O$ values and increases in gypsum precipitation signifying elevated sulfur content. Wind-borne particle concentrations from the annually dated Quelccaya ice core in the Peruvian altiplano are also shown (deMenocal 2001, Fig. 6; American Association for Advancement of Science, by permission)

than northern hemisphere data. From the above, civilizations depended critically on specific climate conditions and have collapsed to a major extent due to natural climate variability which led to economic and social crises, decline in population and war (Diamond 2005; Morris 2011). Collapse may have occurred intermittently, as in Egypt and China, or became terminal as in Easter Island, the Maya.

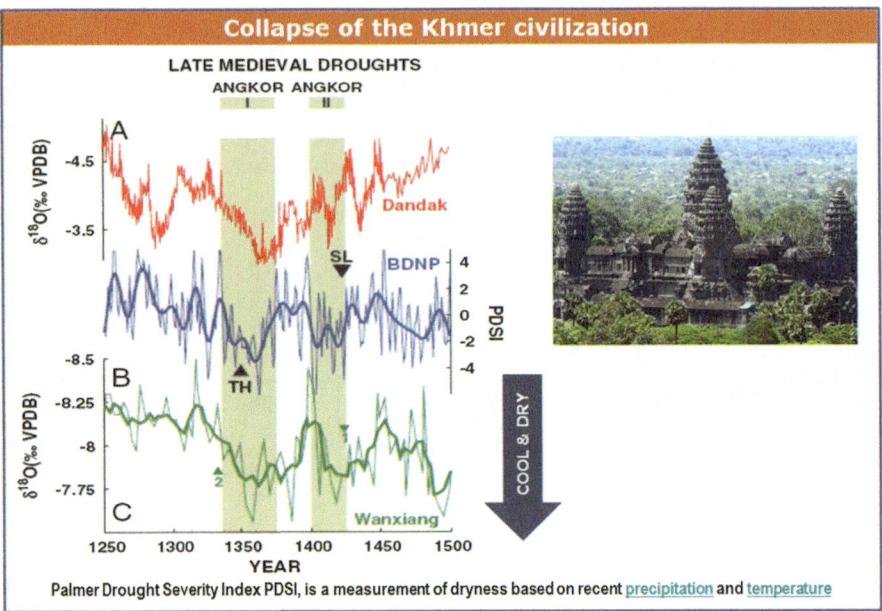

Fig. 7.9 Regional paleoclimate records of medieval drought in Southeast Asia. Dandak Cave $\delta^{18}O$ record (**a**) from the core monsoon region of India and Bidoup Nui Ba National Park (BDNP) PDSI reconstruction (**b** with heavy line 15-year Butterworth filter from southern Vietnam), and the speleothem $\delta^{18}O$ record from Wanxiang Cave (**c** heavy line, five-point boxcar filter) in China. The fourteenth and early fifteenth century Angkor droughts are indicated by the brown shaded bars. Historical records of the fourteenth and fifteenth century droughts come from Phitsanulok in modern Thailand (TH) and Sri Lanka (SL) and are indicated by black triangles (Buckley et al. 2010, Fig. 3. Proceedings National Academy of Science USA, by permission). Photo of Angkor Wat; author: Termer https://commons.wikimedia.org/wiki/File:Angkor_Wat_01.jpg

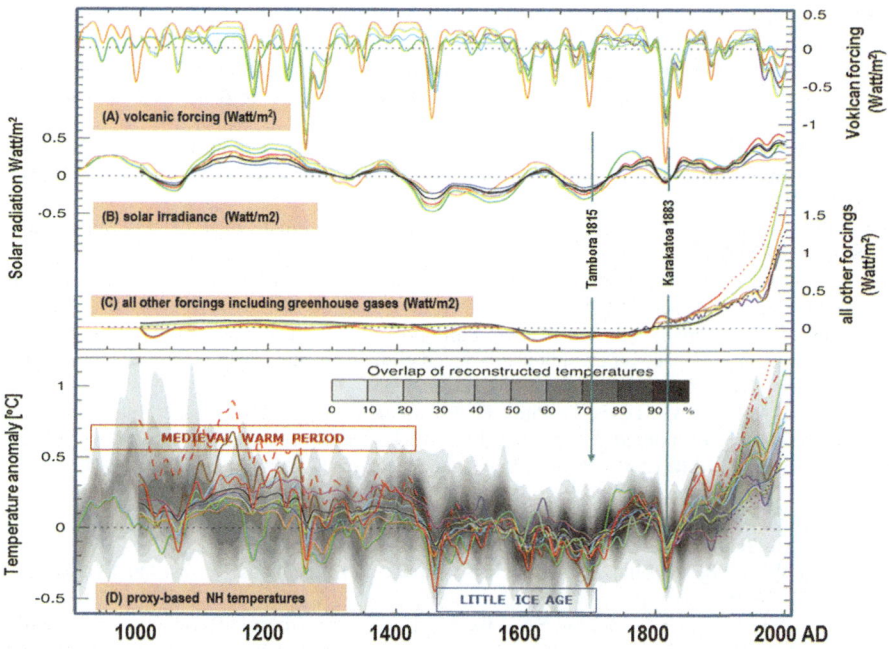

Fig. 7.10 Radiative forcings and simulated temperatures during the last 1.1 kyr. Global mean radiative forcing (Watt/m^2) used to drive climate model simulations due to (**a**) volcanic activity, (**b**) solar irradiance variations and (**c**) all other forcings (which vary between models, but always include greenhouse gases, and, except for those with *dotted lines* after 1900, tropospheric sulphate aerosols). (**d**) Annual mean Northern Hemisphere temperature (°C) simulated under the range of forcings shown in (**a**) to (**c**), compared with the concentration of overlapping NH temperature reconstructions (shown by grey shading, modified to account for the 1500–1899 reference period. All forcings and temperatures are expressed as anomalies from their 1500 to 1899 means and then smoothed with a Gaussian-weighted filter to remove fluctuations on time scales less than 30 years; smoothed values are obtained up to both ends of each record by extending the records with the mean of the adjacent existing values (IPCC-AR4 2007. The Physical Science Basis. Working Group I Contribution to the Fourth Assessment Report of the Intergovernmental Panel on Climate Change. Cambridge University Press. Figure 6.13, by permission http://www.ipcc.ch/publications_and_data/ar4/wg1/en/figure-6-13.html)

References

Broecker WC, Stocker TF (2006) The Holocene CO2 rise: anthropogenic or natural? Eos 87:27–29

Buckley BM, Anchukaitisa KJ, Penny D, Fletcher R, Cook ER, Sano M, Nam LC, Wichienkeeo A, Minh TT, Mai Hong T (2010) Climate as a contributing factor in the demise of Angkor, Cambodia. Proc Nat Acad Sci 107:6478–6752

Clift PD et al (2007) Holocene erosion of the Lesser Himalaya triggered by intensified summer monsoon. Geology 36:79–82

Cullen HM, deMenocal PB, Hemming S et al (2000) Climate change and the collapse of the Akkadian empire: evidence from the deep sea. Geology 28:379–382 deMenocal PB (2001) Cultural responses to climate change during the late holocene. Science 292:667–673

Diamond J (2005) Collapse: how societies choose to fail or succeed. Viking, New York

Gustav Doré (1866) Wikipedia Commons http://en.wikipedia.org/wiki/File:Confusion_of_Tongues.png

IPCC-AR4 (2007) The physical science basis. Working group I contribution to the fourth assessment report of the intergovernmental panel on climate change. Cambridge University Press

Jones PD, Osborn TJ, Briffa KR (2001) The evolution of climate over the last millennium. Science 292:662–667

Jouzel J et al (2007) Orbital and millennial Antarctic climate variability over the past 800,000 years. Science 317:793–797

Kutzbach JE, Ruddiman WF, Vavrus SJ, Philippon G (2010) Climate model simulation of anthropogenic influence on greenhouse-induced climate change (early agriculture to modern): the role of ocean feedbacks. Climatic Change 99:351–381

Lewis CFM, Miller AAL, Levac E, Piper DJW, Sonnichsen GV (2012) Lake Agassiz outburst age and routing by Labrador current and the 8.2 ka cold event. Quatern Int 260:83–97

Lisiecki LE, Raymo ME (2005) A Plio-Pleistocene stack of 57 globally distributed benthic $\delta^{18}O$ records. Paleoceanography 20:PA1003. doi:10.1029/2004PA001071

Morris I (2011) Why the west rules for now: the patterns of history, and what they reveal about the future. Picador, New York

Ruddiman WF (2003) Orbital insolation, ice volume, and greenhouse gases. Quatern Sci Rev 22:1597–1629

Shakun JD, Clark PU, He F, Marcott SA, Mix AC, Liu Z, Otto-Bliesner B, Schmittner A, Bard E (2012) Global warming preceded by increasing carbon dioxide concentrations during the last deglaciation. Nature 484:49–55

Shen H (2012) Drought hastened Maya decline: a prolonged dry period contributed to civilization's collapse. Nature. http://www.nature.com/news/drought-hastened-maya-decline-1.11780

Solanki SK (2002) Solar variability and climate change: is there a link? Sol Phys 43:59–513

Steffensen JP et al (2008) High-resolution Greenland ice core data show abrupt climate change happens in few years. Science 321:680–684

Wagner F, Aaby B, Visscher H (2002) Rapid atmospheric CO2 changes associated with the 8,200-years-B.P. cooling event. Proc Nat Acad Sci 99:12011–12014

Weiss H et al (1993) The genesis and collapse of third millennium north Mesopotamian civilization. Science 261:995–1004

Wiersma AP, Roche DM, Renssen H (2011) Fingerprinting the 8.2 ka event climate response in a coupled climate model. J Quater Sci 26:118–125

Part IV
The Anthropocene Event Horizon

I met a traveler from an antique land
Who said: "Two vast and trunkless legs of stone
Stand in the desert. Near them on the sand,
Half sunk, a shattered visage lies, whose frown
And wrinkled lip and sneer of cold command
Tell that its sculptor well those passions read
Which yet survive, stamped on these lifeless things,
The hand that mocked them and the heart that fed.
And on the pedestal these words appear:
'My name is Ozymandias, King of Kings:
Look on my works, ye mighty, and despair!'
Nothing beside remains. Round the decay
Of that colossal wreck, boundless and bare,
The lone and level sands stretch far away".

Ozymandias

Percy Bysshe Shelley

Chapter 8
Homo sapiens' War Against Nature

Abstract Nature includes species whose activities are capable of devastating habitats, examples include toxic viruses, methane (CH_4) and hydrogen sulphide (H_2S)-emitting bacteria, fire ant armies, locust swarms and rabbit populations. Parasitic host-destroying organisms include species of fungi, worms, arthropods, annelids and vertebrates, cf. oxpeckers and vampire bats. The mastery of fire has enabled the genus Homo to magnify its potential to harness and release energy by orders of magnitude, increasing entropy in nature on a scale unprecedented in the Cenozoic (since 65 Ma). From the mid-20th century, the splitting of the atom allowed humans to trigger a chain reaction potentially devastating much of the biosphere. Once a species has developed sources of energy of this magnitude the species would need to be perfectly wise and responsible if it is to prevent its inventions to get out of control.

8.1 Neolithic Burning and Early Global Warming

Proceeding from pre-historic burning, such as fire-stick farming by the Australian aborigines (Gammage 2012), the dawn of the Neolithic about ~10,000 years-ago saw an expansion of anthropogenic burning, clearing land for cultivation of crops. Pottery, smelting of metals (iron, bronze, copper, and gold), possibly discovered accidentally around camp fires, have led to crafting of ploughs to till the land and swords to kill enemies. Extensive burning and land clearing during the Holocene further magnified entropy. During this period biomass burning, as indicated by residual charcoal deposits, has reached levels possibly as high as those resulting from the combustion of fossil fuels during the first part of the 20th century (Bowman et al. 2009) (Fig. 5.2).

The use of fire for clearing land for farming has been central to Neolithic culture. North American Indians conducted deliberate seasonal or periodic burnings aimed at forming mosaics of resource diversity, environmental stability and the maintenance of transition zones (ecotones) (Lewis 1985). This was compounded by deliberate and incidental burnings spreading out of camp fires, hunting, smoke signals and inter-tribal wars. Controlled surface burns were broken by occasional escape of wild fires and periodic conflagrations during times of drought (Pyne 1982, 1995). So extensive were the cumulative outcomes of these fires that the overall effect of Indian occupation of the Americas replaced extensive forested land with grassland or savannah, or, where the forest persisted, opened it up and freed it from underbrush.

Major studies of North American fire regimes (Lewis 1985; Kay 1994; Russell 1983) list factors underlying deliberate and accidental ignition, including:

1. Burning aimed at diversion of big game (deer, elk, bison) for hunting, opening of prairie and meadows for host pasture and grazing and of small areas to trap other game (rabbits, raccoons, bears. ducks, geese).
2. Burning to clear areas for planting of crops, like corn and tobacco, facilitate grass seed, berry and medicine plant collection, prevent abandoned fields from overgrowth, clear grass and brush to facilitate the gathering of acorns, roast mescal and obtain salt from grasses.
3. Forming fire enclaves for collection of insects (crickets, grasshoppers, moths) and collection of honey from bees.
4. Back-burning for protection of settlements from wild fires.
5. Burning for protection from pests (rodents, snakes, black flies, mosquitos).
6. War and signaling, including burning strategies against enemies.
7. Burning and girdling of trees for firewood and timber.
8. Clearing undergrowth for access and travel.
9. Worship and spiritual rain dances.

Whereas burning practices by North American Indians and other people (Sheuyangea et al. 2005) were primarily aimed at land clearing for agriculture and or hunting, the rise of agricultural civilizations capable of supporting armies unleashed scorched earth strategies on a massive scale. Examples include the razing of circum-Mediterranean forests and agricultural lands by retreating armies, for example recorded by Gibbon (1788) in connection with the advance of the Emperor Julian into Mesopotamia:

> … on the approach of the Romans, the rich and smiling prospect was instantly blasted. Wherever they moved… the cattle was driven away; the grass and ripe corn were consumed with fire; and, as soon as the flames had subsided which interrupted the March of Julian, he beheld the melancholy face of a smoking and naked desert.

Ruddiman (2003) defines the onset of a human-dominated Anthropocene era on the basis of a rise in CO_2 from about $\sim 6{,}000$ years-ago and of methane from about $\sim 4{,}000$ years-ago, consequent on land clearing, fires and rice cultivation. Kutzbach et al. (2010), comparing Holocene temperature variations with those of earlier interglacial periods, estimates the rise of anthropogenic greenhouse gas levels during the Holocene prevented a decline in temperatures into the next glacial cycle by as much as ~ 2.7 °C (Fig 7.6). By contrast Crutzen and Stoermer (2000) and Steffen et al. (2007) define the onset of the Anthropocene at the dawn of the industrial age in the 18th century. According to the latter classification the mid-Holocene rise of CO_2 and methane was related to a natural trend, as based on comparisons with the 420–405 kyr Holsteinian interglacial (Broecker and Stocker 2006). Other arguments which favor a non-anthropogenic rise in greenhouse gases during the mid-Holocene hinge on CO_2 mass balance calculations (Stocker et al. 2010), CO_2 ocean sequestration rates and calcite compensation depth (Joos et al. 2004). However, while the signatures of pre-industrial anthropogenic emissions and natural variability are difficult to discriminate, there can be little doubt human-triggered fires and land clearing contributed to an increase in greenhouse gases through much of the Holocene.

8.2 The Great Carbon Oxidation Event

Beginning with the 18th century, the onset of combustion of coal, oil and gas which underpin the industrial age has led to the release of near > 360 GtC (billion ton carbon), compounded by land clearing and fires which led to the release of near ~ 150 GtC to the atmosphere (Global Carbon Project), the added near-560 GtC being just under the pre-industrial atmospheric carbon level of ~ 590 GtC, with major consequences for atmospheric CO_2 levels (Figs. 8.1–8.9). Emissions of both CO_2 and SO_2 have grown during WWI and WWII (Figs. 8.1 and 8.6), the largest

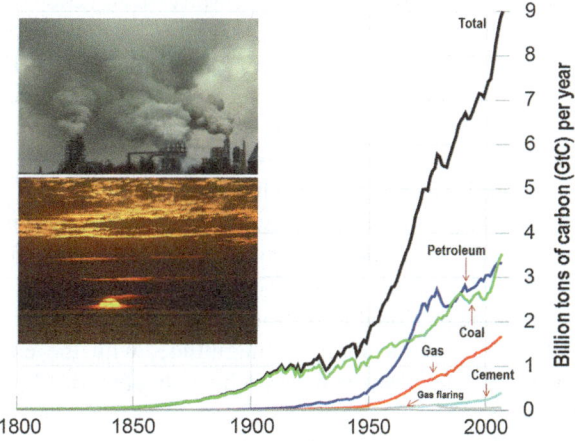

Fig. 8.1 Global CO_2 emission estimate since 1800 AD. Carbon Dioxide Information Analysis Center Oak Ridge National Laboratory http://cdiac.ornl.gov/trends/emis/glo.html. (courtesy Tom Boden, with permission). Inset 1: Coal power plant in Datteln (Germany) at the Dortmund-Ems-Kanal; Image by Arnold Paul cropped by Gralo http://en.wikipedia.org/wiki/File:Coal_power_plant_Datteln_2_Crop1.png ; Inset 2: sunset at Porto Covo, west coast of Portugal. Author: Joaquim Alves Gaspar, Lisboa, Portugal. Wikipedia Commons http://en.wikipedia.org/wiki/File:Sunset_2007-1.jpg

acceleration occurs since about 1950, reflecting the post-war economic boom (Fig. 8.1). Combined with other greenhouse gases this has led to an increase in the radiative forcing of the atmosphere by about +3.05 Watt/m^2. Approximately −1.2 Watt/m^2 is masked by direct and indirect (cloud-related) effects of emitted sulphur aerosols (IPCC 2007; Hansen et al. (2012a) (Fig. 8.2). Emissions continued to rise during 1978–2011, including acceleration of CO_2, a temporary halt of methane rise during 1996–2002 resumed in 2007. By 2010 about 9 billion tons of carbon have been annually released to the atmosphere (Fig. 8.1), of which near-42 % of emitted CO_2 stay in the atmosphere, near to one quarter is absorbed by the oceans and near to one quarter is sequestered by vegetation and soils on land (Figs 8.4 and 8.5). Between March 2012 and March 2013 an unprecedented rise from 394.45 to 397.34 ppm has taken place (NOAA 2013).

The Australian CSIRO reports a rise of atmospheric CO_2 concentration from ∼295 to ∼380 ppm, of CO_2-e (combined CO_2 and methane equivalent) from ∼300 to ∼460 ppm, and radiative forcing as more than 2.2 Watt/m^2 during 1900–2011. A mean global temperature lull and a mild degree of cooling during ∼1940–1970, related to rising levels of sulphur aerosols and to a low in the 11 years sun spot cycle, were abruptly terminated by a warming trend from 1975 when atmospheric CO_2 level reached ∼330 ppm and clear air policies reduced sulphur emissions (Fig. 8.6). SO_2-induced slowing-down of the rate of warming also occurs from about 2001, related to a rise in sulphur emissions from China (Smith et al. 2011) (Figs. 8.6 and 8.7). The direct and indirect negative forcing effect of sulphur aerosols since the onset of the industrial age, masking global warming, is estimated as −1.2 Watt/m^2 by IPCC (2007), −1.5 Watt/m^2 by Hansen et al. (2012a) and −1.1 ± 0.4 Watt/m^2 by Murphy et al. (2009).

Measured mean global temperature rise of approximately +0.8 °C since 1800 AD are amplified in polar regions to + 4–5 °C (NASA/GISS 2013), representing the effects of rising polar sea water temperatures, related decrease in the effect of ice albedo reflection and increase in open water infrared absorption—namely an albedo-flip feedback process leading to melting of continental ice sheets (Fig. 8.10). The overall increase in radiative forcing in the atmosphere of ∼3.17 Watt/m^2 (Fig. 8.2) is near ∼50 % that involved in the last glacial termination, where loss of ice cover and vegetation accounted for + 3.5 ± 1.0 Watt/m^2 and rise in GHG for +3.0 ± 0.5 Watt/m^2 (Hansen et al. 2008). The rise rates of greenhouse gases and mean global temperatures exceed that of recorded Cenozoic climate shifts by more than an order of magnitude, with the exception of temperature rise rates of the Dansgaard-Oeschger cycles of the last glaciation (Fig. 8.9) (Table A2).

The atmospheric CO_2 rise from ∼280 to 397–400 ppm, with a mean of 0.43 ppm/year and reaching 2.9 ppm/year in 1998 and ∼2.89 ppm/year during 2012–2013 (NOAA 2013), exceeds any measured in the geological record, including the PETM hyperthermal methane-release event and the K-T asteroid impact (Zachos et al. 2008; Beerling et al. 2002) (Fig. 8.9; Table A2). Measured temperature rise rates of ∼0.003 °C/year (or without aerosol masking by 0.008C/year since 1750) (Table A2) exceed all previous rates, excepting the Dansgaard-Oeschger (D-O) rates in the range

8.2 The Great Carbon Oxidation Event

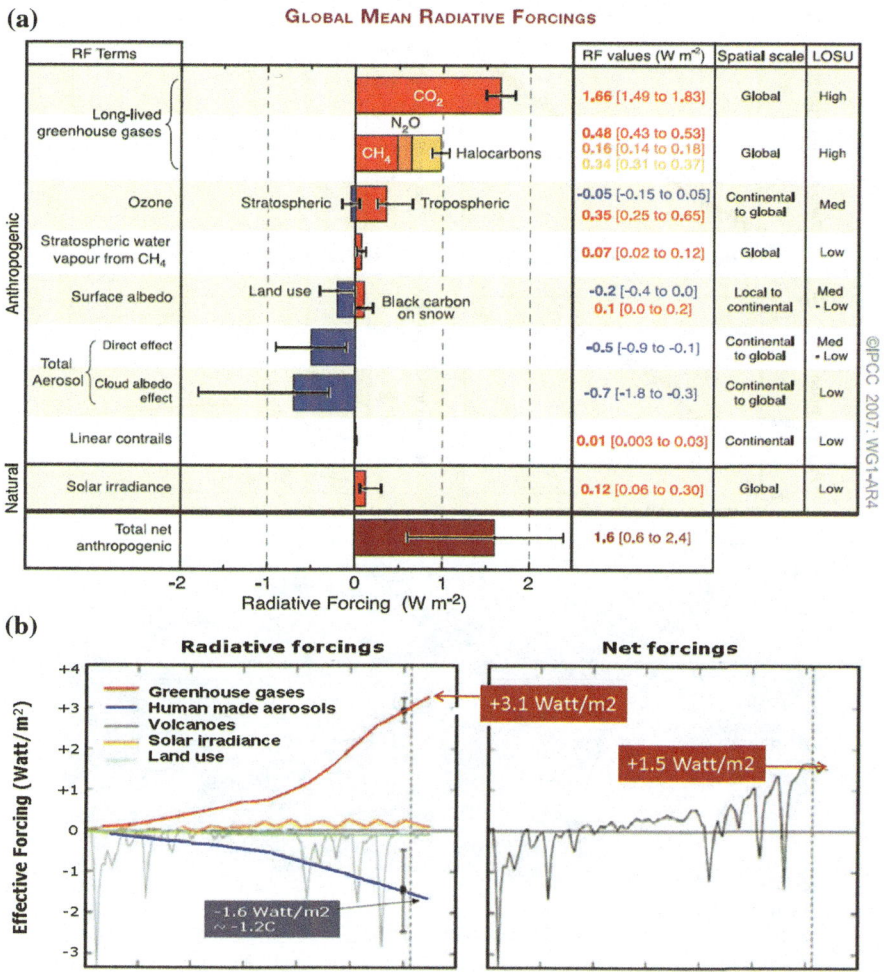

Fig. 8.2 a Global mean positive and negative radiative forcings since 1750. Global average radiative forcing estimates and ranges in 2005 for anthropogenic carbon dioxide (CO_2), methane (CH_4), nitrous oxide (N_2O) and other important agents and mechanisms, together with the typical geographical extent of the forcing and the assessed level of scientific understanding. The net anthropogenic radiative forcing and its range are also shown. Volcanic aerosols contribute an additional natural forcing but are not included in this figure due to their episodic nature (IPCC-AR4 2007. The Physical Science Basis. Working Group I Contribution to the Fourth Assessment Report of the Intergovernmental Panel on Climate Change. Cambridge University Press. figure TS.5, by permission). http://www.ipcc.ch/publications_and_data/ar4/wg1/en/tssts-2-5.html).
b Climate forcings through 2003 (*vertical line*) and aerosol forcing after 1990, approximated −0.5 times the GHG forcing. Aerosol forcing includes all aerosol effects, including indirect effects on clouds and snow albedo. GHGs include O_3 and stratospheric H_2O, in addition to well-mixed GHGs. (Hansen et al. 2012a, b Fig. 1; Courtesy J. Hansen)

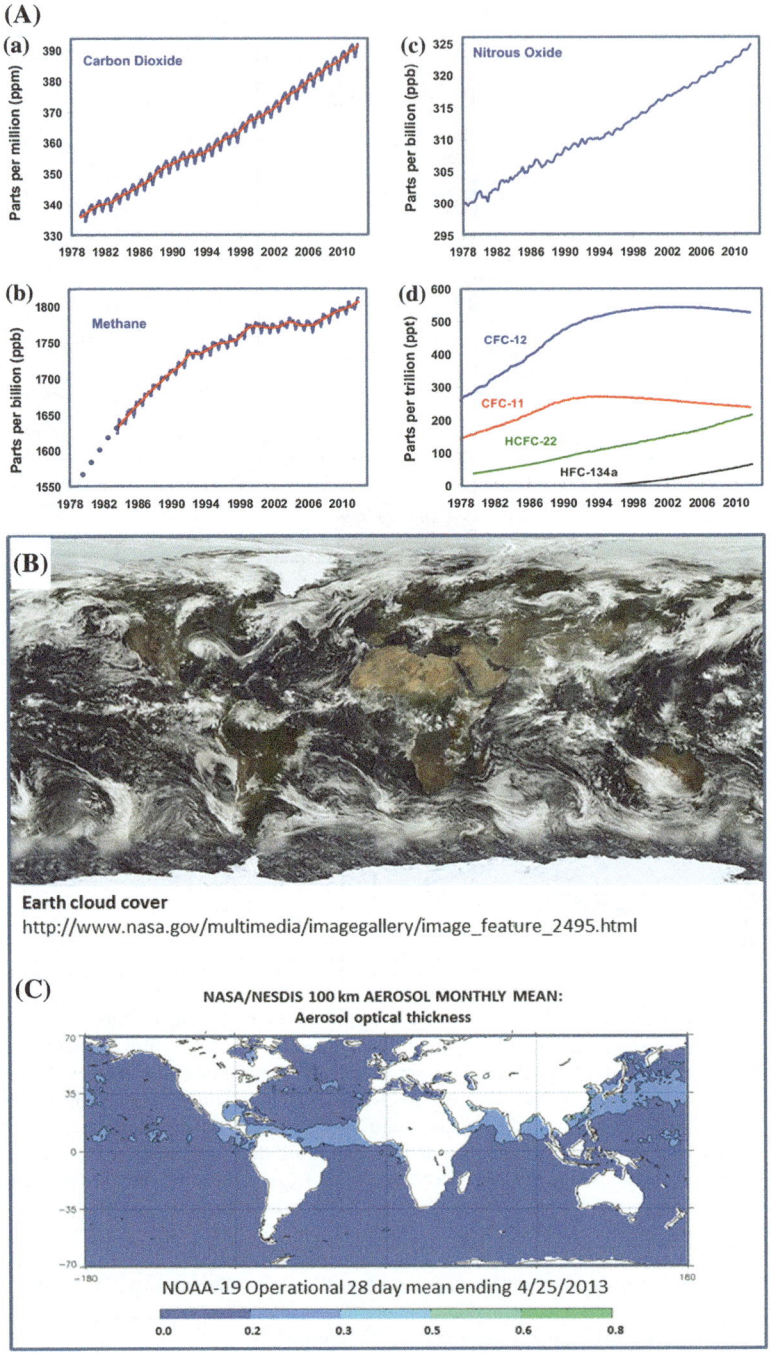

8.2 The Great Carbon Oxidation Event

Fig. 8.3 Trends in the atmosphere. **A** (**a**) CO_2; (**b**) methane; (**c**) Nitrous oxide, and (**d**) CFC and HFC (NOAA)—Global average abundances of the major, well-mixed, long-lived greenhouse gases-carbon dioxide, methane, nitrous oxide, from the NOAA global air sampling network are plotted since the beginning of 1979. These gases account for about 96 % of the direct radiative forcing by long-lived greenhouse gases since 1750 (http://www.esrl.noaa.gov/gmd/aggi/); **B** Earth image displaying cloud cover, Terra satellite Moderate Resolution Imaging Spectroradiometer (MODIS) sensor. July 11, 2005. http://www.nasa.gov/multimedia/imagegallery/image_feature_2495.html **C** NOAA aerosol optical depth for the month ending 25.4.2013 over oceanic regions. Note the concentration of aerosols in tropical zones as well as off northeast Asia. http://www.ospo.noaa.gov/data/aerosol/aermon.fc.gif

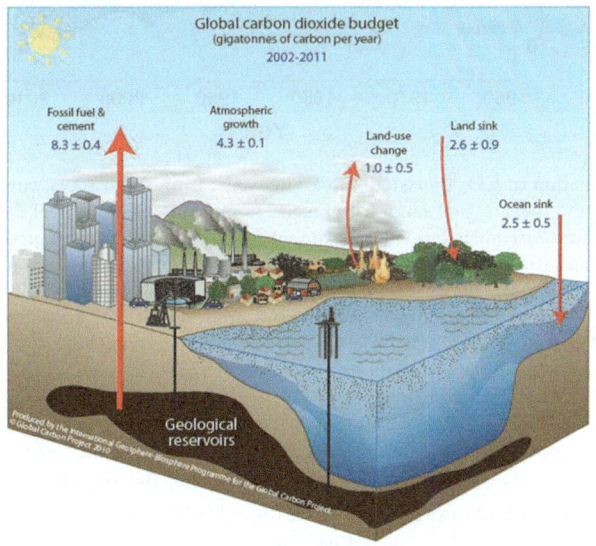

Fig. 8.4 The global carbon budget 2002–2011 in Gigaton carbon per-year (GtC/year). Schematic representation of the overall perturbation of the global carbon cycle caused by anthropogenic activities, averaged globally for the decade 2002–2011. The arrows represent emission from fossil fuel burning and cement production; emissions from deforestation and other land-use change; and the carbon sinks from the atmosphere to the ocean and land reservoirs. The annual growth of carbon dioxide in the atmosphere is also shown. All fluxes are in units of GtCyr−1, with uncertainties reported as ±1 sigma (68 % confidence). (Le Quéré et al. 2012, Global Carbon Project, Fig. 1; Courtesy P. Canadell). http://www.earth-syst-sci-data-discuss.net/5/1107/2012/essdd-5-1107-2012.pdf

of > 0.1 °C/year (Fig. 8.9). A rise of CO_2-equivalent above 500 ppm and mean global temperatures above + 4 °C would track toward greenhouse Earth conditions such as existed during the early Eocene some 50 million years-ago (Zachos et al. 2001). Based on paleoclimate studies, the current levels of CO_2 of 397.34 ppm (Mouna Loa CO_2, 2013) and of CO_2-equivalent of above > 455 ppm (IPCC-2007), a value which includes the equivalent effects of methane and nitrous oxide, commit the atmosphere to a warming trend approaching the upper stability level of the Antarctic ice sheet of 500 ± 50 ppm CO_2.

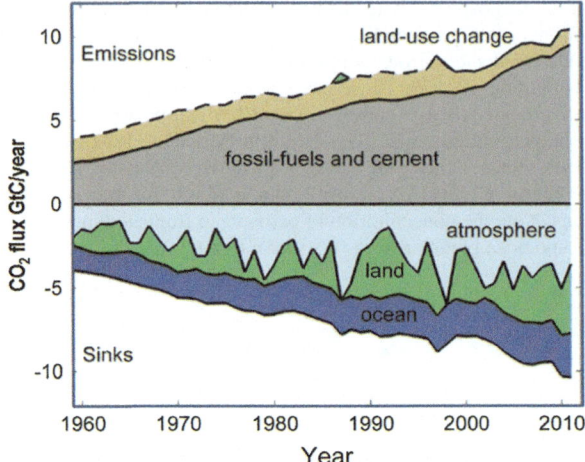

Fig. 8.5 Fractionation of CO_2 emissions (flux GtC/year) from fossil fuels, cement and land use between atmosphere, oceans and land. Combined components of the global carbon budget as a function of time, for (*top*) emissions from fossil fuel combustion and cement production (*grey*) and emissions from land-use change (*brown*), and (*bottom*) their partitioning among the atmosphere (*light blue*), land (*green*) and oceans (*dark blue*). All time-series are in GtC yr−1. Land-use change emissions include management-climate interactions from year 1997 onwards, where the line changes from dashed to full. (Le Quéré et al. 2012, Global Carbon Project, Fig. 2; Courtesy P. Canadell) http://www.earth-syst-sci-data-discuss.net/5/1107/2012/essdd-5-1107-2012.pdf

Based on palaeoclimate studies, using multiple proxies, including soil carbonate $\delta^{13}C$, alkenones, boron/calcium, stomata leaf pores, the current levels of CO_2 of 396–400 ppm and of CO_2-equivalent (a value which includes the equivalent effects of methane and nitrous oxide) of >470 ppm commit the atmosphere to a warming trend tracking toward ice-free Earth conditions (Figs 2.1a, b; 3.1a, b; 3.2).

Global warming is expressed by expansion of the tropics and a shift of high-pressure mid-latitude high pressure ridges toward the poles, where warming (Fig. 8.8a, b) leads to ice/water albedo-flip inherent in melting of sea ice and ice sheets (Fig. 8.10). The rise in sea water temperature enhances the hydrological cycle, with consequent floods and hurricanes. Current trends are shifting the atmosphere to a state analogous to the end-Pliocene, before 2.8 Ma-ago, a period when the bulk of the Greenland and west Antarctic ice sheets melted (PRISM 2012). The Earth's polar ice caps, source of cold air vortices and ocean currents such as the Humboldt and California current, which keep the Earth's overall temperature in balance (Fig. 8.16), are melting at an accelerated rate (Rignot et al. 2011). Melting of the large ice sheets is doubling every 5–10 years (Fig. 8.10), facilitated by sub-glacial melt flow (Chandler et al. 2013) and exceeding the contribution to sea level rise from thermal expansion and mountain glaciers. In Greenland mass loss increased from 137 GtI/year in 2002–2003 to 286 GtI/year in 2007–2009, i.e., an acceleration of -30 ± 11 Gt/year^2 in 2002–2009. In Antarctica mass loss increased

8.2 The Great Carbon Oxidation Event

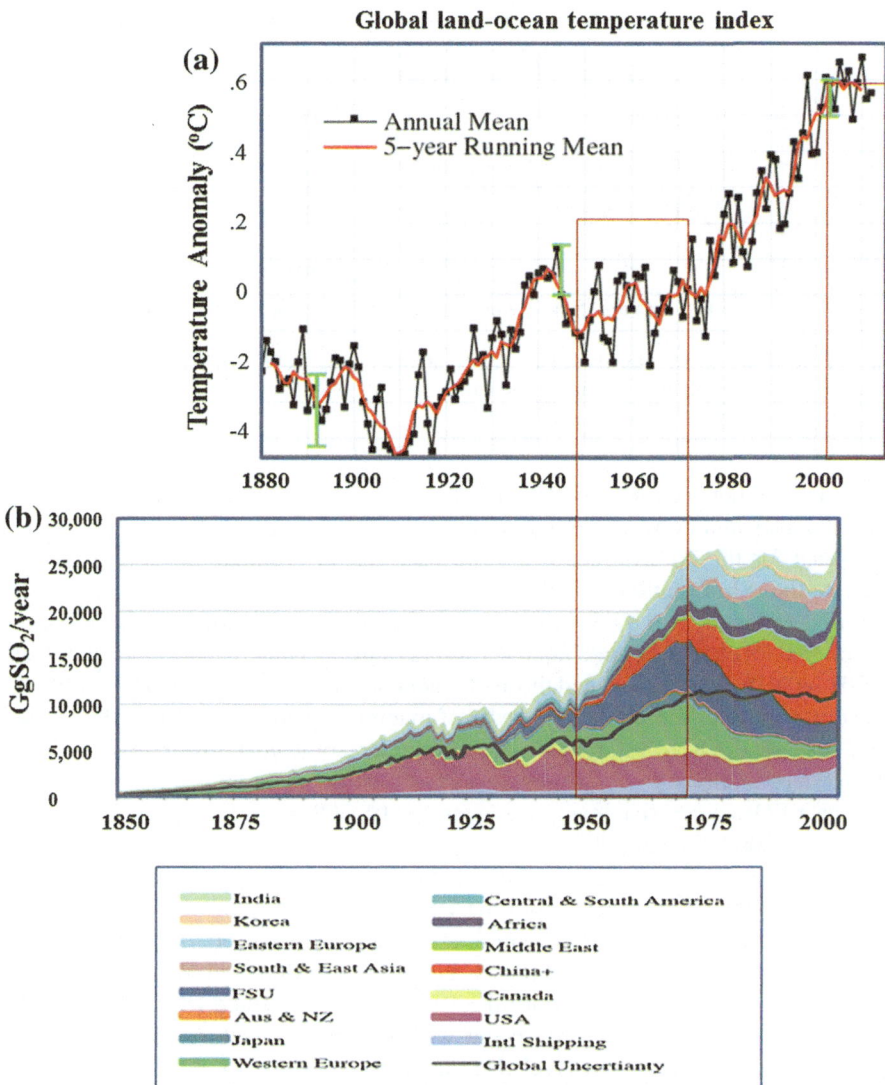

Fig. 8.6 Effects of sulphur emissions on global temperatures. **a** Annual mean and 5 years running mean temperatures 1800–2010 (NASA GISS/Temperature. http://data.giss.nasa.gov/gistemp/graphs_v3/; **b** Global SO2 emissions 1850–2005 (Smith et al. 2011, Fig. 4, Atmos. Chem. Phys., 11:1101–1116, by permission). Note the overlap between increased emission of SO_2 and the slow-down in warming during ~1950–1975 and from about ~2001

from 104 GtI/year in 2002–2006 to 246 GtI/year in 2006–2009, i.e., an acceleration of -26 ± 14 GtI/year^2 in 2002–2009 (Velicogna 2009).

Sea level rise sensitively represents the sum-total of climate change processes, including thermal expansion of water, melting ice sheets and mountain glaciers.

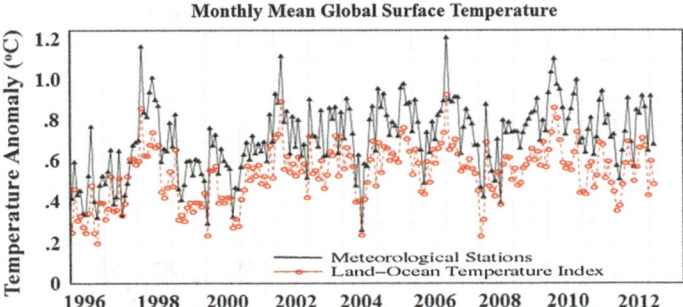

Fig. 8.7 Monthly mean global surface temperatures during 2006–2012 including variations recorded in meteorological stations and from satellite (NASA/GISS Temperatures. http://data.giss.nasa.gov/gistemp/graphs_v3/Fig.C.gif)

The relations between sea levels and mean temperatures during the glacial-interglacial cycles and between temperature rise rate (T°C/year) and sea level rise rate (SL/year) for the 20–21st centuries and past glacial terminations (Fig. 8.11a, b) indicate the (SL/T) ratio of glacial terminations ($\sim$7–20) exceed those of 20–21st centuries ratios ($\ll$1.0) by one to two orders of magnitude, giving a measure of the lag of sea level rise behind land-sea temperatures. Thus, (1) Modern T°C/year (0.0034–0.016 °C/year) exceed those of glacial terminations ($\sim$0.0001–0.0004 °C/year) by near to or more than one order of magnitude (Fig. 8.11b); (2) Modern sea level rise rates ($\sim$1–5 mm/year) are similar or somewhat lower than those of glacial terminations ($\sim$3–10 mm/year), and both are significantly lower than Dansgaard-Oeschger sea level rise rates ($\sim$10–300 mm/year) (Ganopolski and Rahmstorf 2002; Jouzel 2005).

Kemp et al. (2011) reconstructed sea levels for the past 2,100 years from salt-marsh sedimentary sequences from the US Atlantic coast, showing sea level was (1) stable from at least 100 BC until 950 AD; (2) increased from 950 AD for 400 years at a rate of 0.6 mm/year as related to the Medieval Warm Period; (3) was stable or slightly declined until the 19th century; (4) since 1865–1892 rose by an average of 2.1 mm/year, consistent with global temperature. Since the early 20th century, the rate of sea level rise increased from $\sim$1 to $\sim$3.5 mm/year, the 1993–2009 rate being 3.2 ± 0.4 mm/year, a three to four-fold increase since the onset of the industrial age (Rahmstorf 2007) (Fig. 8.12).

With ensuing desertification of temperate zones, e.g. North Africa, southern Europe, south and southwest Australia and southern Africa, forests become prey to heat waves and firestorms (Hansen et al. 2012a, b). Warming of the oceans leads to a decrease in CO_2 solubility, lowered pH and decrease in biological calcification. Increased evaporation in warming oceans leads to an enhanced hydrological cycle, including abrupt precipitation events, floods, and the intensification of cyclones and associated destruction of vegetation (Munich Re-Insurance 2012). Hansen et al. (2012a) estimate the current Earth radiative imbalance, namely of heat trapped in the Earth surface and atmosphere, as $\sim$0.6 Watt/m^2. According to

8.2 The Great Carbon Oxidation Event

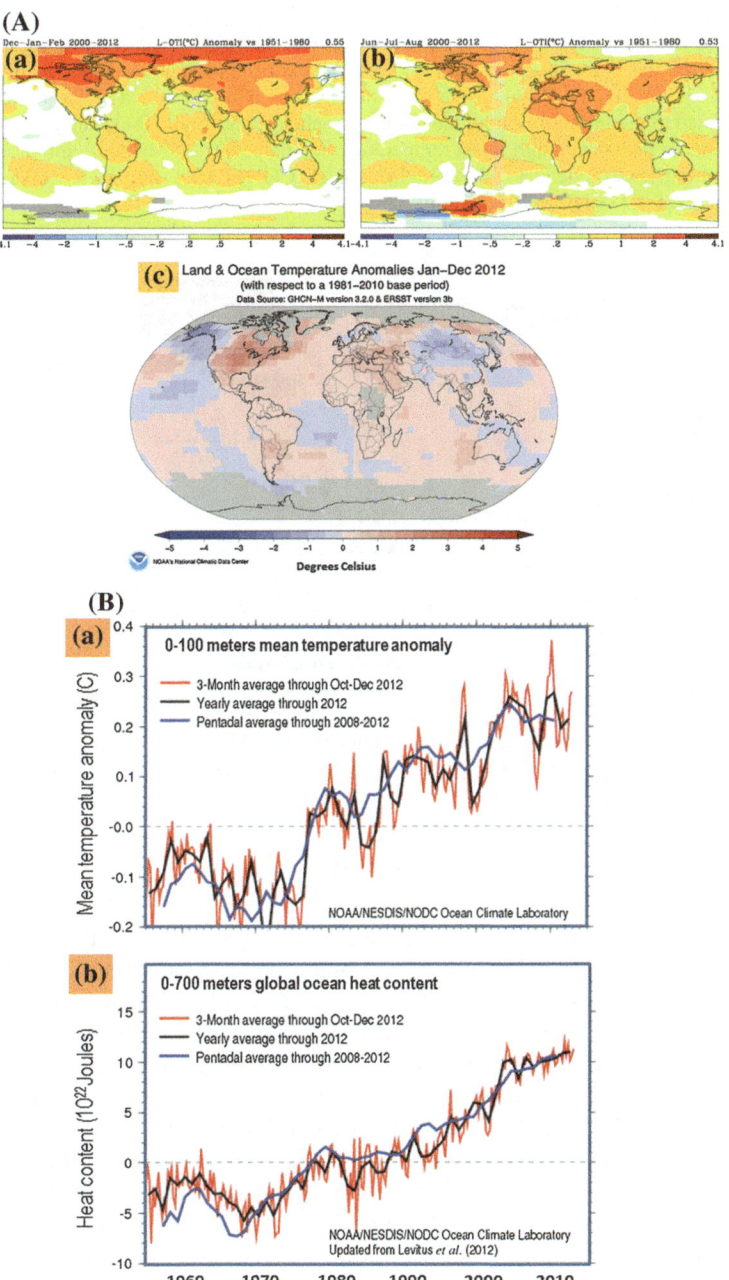

Fig. 8.8 A Global warming (**a**) during Dec–Feb 2000–2012 relative to 1951–1980; **b** during Jun–August 200–2012 relative to 1951–1980, and (**c**) Jan–Dec 2012 relative to 1981–2010. *Sources* http://data.giss.nasa.gov/gistemp; http://www.ncdc.noaa.gov/sotc/service/global/map-blended-mntp/201201-201212.gif Note the strong rates of warming in polar regions during winter, related to retention of heat by the higher greenhouse gas levels. **B** Ocean temperatures and heat contents: **a** 0–100 m depth mean temperature anomaly; **b** 0–700 m depth global ocean heat content. http://www.nodc.noaa.gov/OC5/3M_HEAT_CONTENT/ http://www.nodc.noaa.gov/OC5/3M_HEAT_CONTENT/index3.html

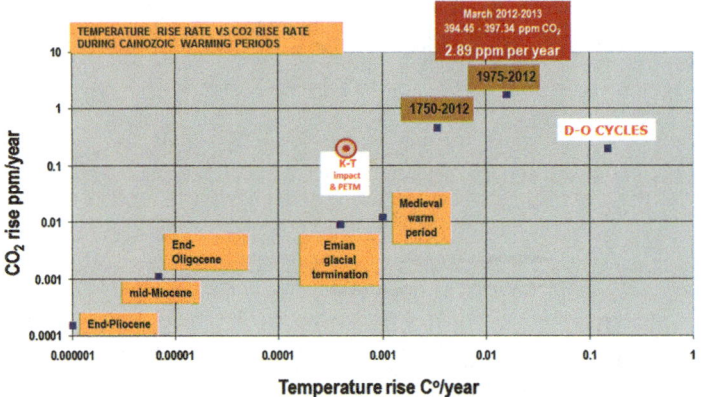

Fig. 8.9 Relations between CO_2 rise rates and mean global temperature rise rates during warming periods, including the Paleocene-Eocene Thermal Maximum, Oligocene, Miocene, late Pliocene, Eemian (glacial termination), Dansgaard-Oeschger cycles, Medieval Warming Period, 1750–2012 and 1975–2012 periods. *Data sources* Tables A1 and A2

Hansen (2012a, b) this equates "400,000 Hiroshima atomic bombs per day 365 days per year".

The onset of an irreversible change in global climate from relatively stable state across critical threshold, referred to as tipping point (Lenton et al. 2009), occurs when the climate system transgresses a point beyond which amplifying positive feedbacks drive the process until negative feedbacks, such as significant decrease in solar insolation or depletion in the source of greenhouse gases, stabilize a new climate state. Such climate shifts may consist of multiple events, including melting and collapse of Greenland and west Antarctic ice sheets, much of the latter being grounded below sea level, melting of permafrost, boreal forest dieback and tundra loss, Indian and west African monsoon shifts, Amazon forest dieback, ozone hole growth and changes to the ENSO circulation and ocean deep water formation patterns (Fig. 9.5). In desert and semi-arid regions global warming leads to heat waves and droughts, which have increased by at least a factor of two since the 1980 (Munich Re-insurance 2012) (Figs. 8.13 and 8.14). Upon drying up temperate regions become prey to firestorms, driving loss of CO_2 from desiccated and burnt vegetation and soils.

8.2 The Great Carbon Oxidation Event

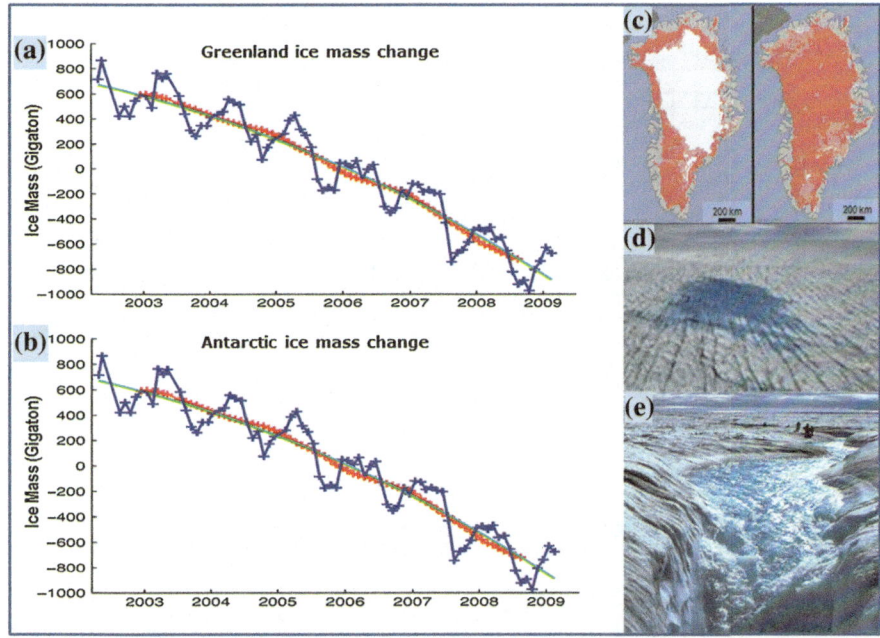

Fig. 8.10 Rates of ice melt in **a** Greenland and **b** Antarctic (Velicogna 2009). Mass change deduced from gravitational field measurements Time series of ice mass changes for the Antarctic ice sheet estimated from GRACE monthly mass solutions for the period from April 2002 to February 2009. Unfiltered data are blue crosses. (Velicogna 2009; Figs. 1 and 2; Geophysical Research Letters, by permission). (**c**) Greenland summer ice extent (NASA) http://www.nasa.gov/topics/earth/features/greenland-melt.html; **d** Greenland surface ice melt (NASA); **e** Moulin drainage conduit, Greenland. Melt water flows into a large moulin in the Greenland ice sheet. Image credit: Roger J. Braithwaite, The University of Manchester, UK. *Image Source*: http://www.wunderground.com/climate/greenland.asp. (NASA)

The rise in temperatures decreases the ocean's ability to sequester CO_2. The dissolved gas forms bicarbonate and carbonic acid (Fig. 8.17), thereby lowering the water pH. In turn this results in dissolution of biogenic carbonates in plankton and corals, retarding CO_2 capture. The deep ocean sink for CO_2 is >39,000 GtC, near to 66 times that of the original CO_2 level of the atmosphere. The oceans absorb nearly 40 % of the anthropogenic CO_2 (Fig. 8.5), leading to an increase in the concentration of carbonic acid (H_2CO_3) and thus an overall increase in acidity (Fig. 8.17), namely lowering of seawater pH. As the mixing time between surface and deep ocean waters is in the order of a 1,000 years, a disequilibrium ensues between the surface and the deep oceans (McCulloch 2008). The carbonate ion CO_3^{-2} concentration of seawater is sensitive to minor changes in pH and a shift from ~8.2 (the current value) to ~8.0 due to doubling of atmospheric CO_2 would reduce CO_3^{-2} by nearly 40 %, with major consequences for calcifying organism (McCulloch 2008), including coral reefs and plankton. Coral reefs are particularly

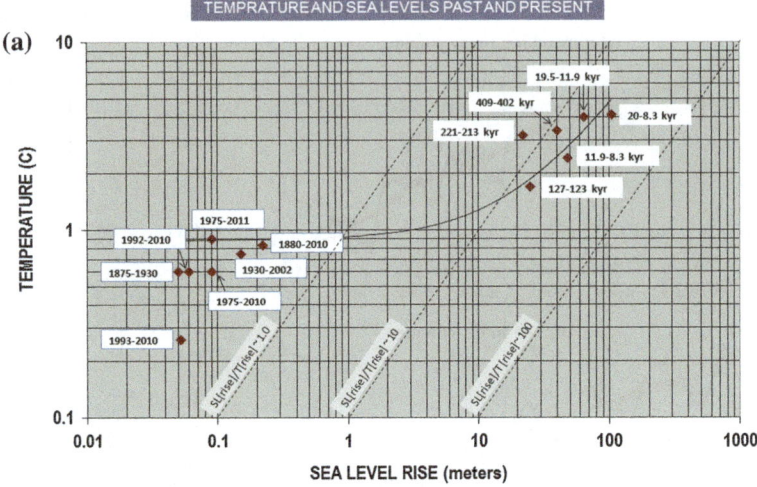

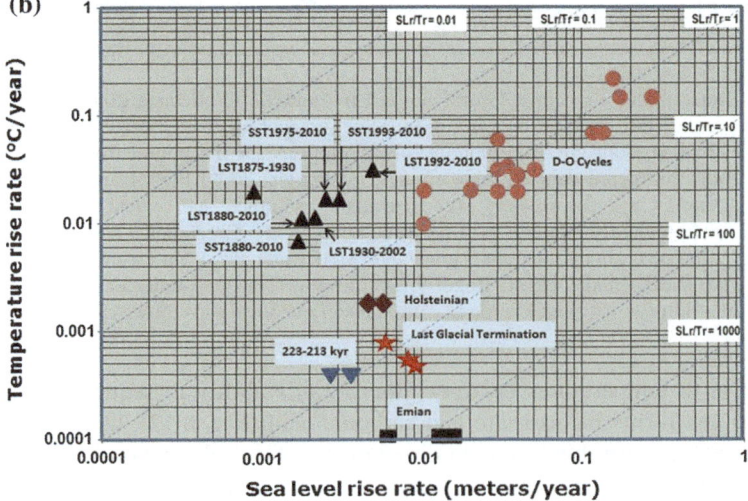

Fig. 8.11 a Relations between sea levels (SL in meters) and temperatures (T°C) during glacial terminations and post-1750 periods. Note the lag by more than an order of magnitude of the ratio of sea level rise to temperature rise, which is mainly >1.0 to >10 for glacial terminations and below <1.0 for post-1750 time intervals. Data sources: Temperatures—Hansen et al. (2007, 2008); IPCC AR4, (2007). Sea levels—IPCC AR4 (2007) and Siddall et al. (2003). **b** Relations between temperature rise rates (T°C/year) and sea level rise rates (meters/year) during glacial termination intervals, Dansgaard-Oeschger (D-O) cycles and post-1750 periods. Modern temperature rise rate exceed glacial termination rates by about or more than an order of magnitude. D-O temperature and sea level rise rates are higher than those of both glacial terminations and modern rates. Modern sea level rise to temperature rise are the lowest, representing lag effects

8.2 The Great Carbon Oxidation Event

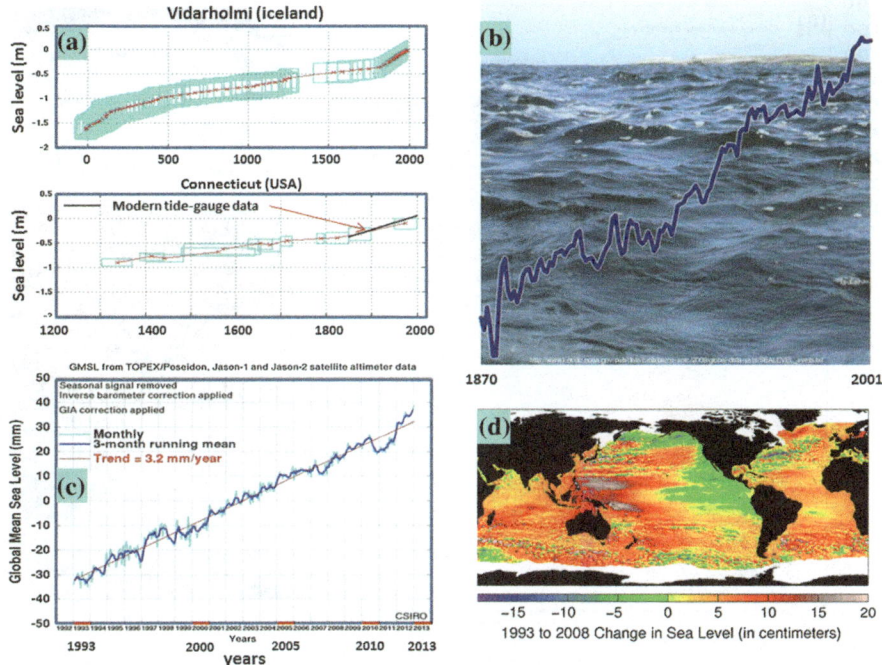

Fig. 8.12 Historic and 20th to 21st centuries sea level changes: **a** Historic sea level changes in Vidarholmi, Iceland and Connecticut; **b** 1870–2001 sea level changes (NOAA); **c** 1993–2013 sea level change rate; **d** 1993–2008 total sea level changes. Note the high sea level changes rates in the west Pacific Ocean, related to the westward flow of warm water (CSIRO sea level changes http://www.cmar.csiro.au/sealevel/sl_hist_last_15.html, by permission)

sensitive—and the large mass extinctions left the Earth without living reefs for at least four million years, referred to as "reef gaps" (Veron 2008). Causes of these extinctions include (1) changes independent of the carbon cycle, including direct physical destruction from asteroid impacts, dust clouds, sea-level regressions during glacial periods, extreme temperature changes, salinity, diseases and toxins; (2) changes linked to the carbon cycle, including acid rain, variations in ocean chemistry, including hydrogen sulphide, anoxia, methane, carbon dioxide and pH. The prospect of ocean acidification has the potential to trigger a sixth mass extinction event in the oceans (Veron 2008).

In a landmark paper *Extinction Risk from Climate Change* (Thomas 2004) review shifts and redistribution of species during the previous 30 years, which form the basis for assessments of future extinction risks for regions covering about 20 % of the land surface. Assuming mid-range climate warming for 2050 some 15–37 % of species in the sample regions are committed to extinction. Minimal climate-warming scenarios result in lower extinctions of about ∼18 % and maximum-change in ∼35 %, namely driving more than one million species to extinction. A comprehensive publication edited by Hannah (2011) *Saving a million*

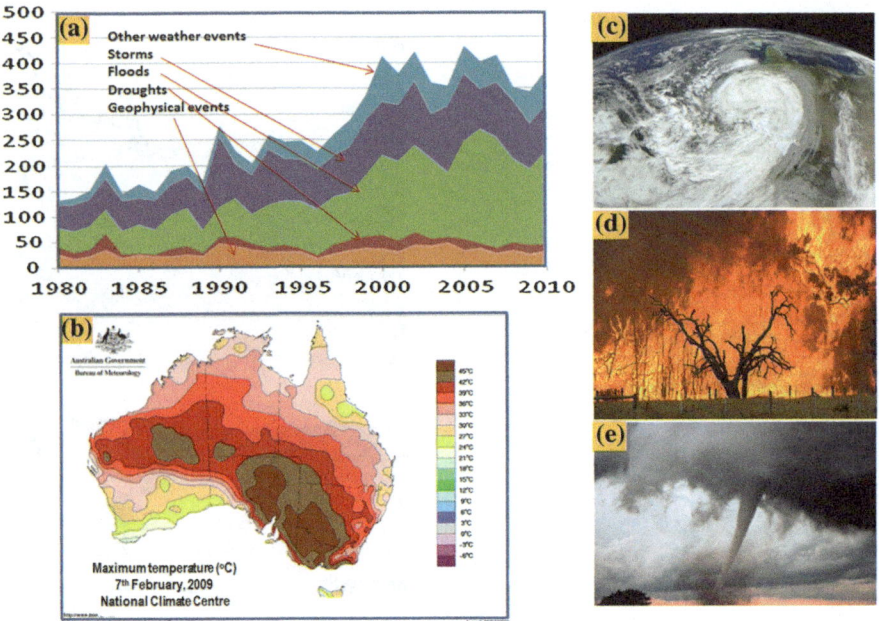

Fig. 8.13 Extreme weather events: **a** Number of reported disasters 1980–2010 (http://make wealthhistory.org/2011/05/30/the-number-of-natural-disasters-is-on-the-rise/) Courtesy Gifty Dzah, Oxfam GB Policy and Practice Communications Team); http://eoimages.gsfc.nasa.gov/images/imagerecords/79000/79607/sandyhalf_vir_2012304_lr; **b** Maximum temperature 7th February, 2009 (Australian Bureau of Meteorology, by permission); **c** Hurricane Katrina (NASA); **d** Fire storm, Victoria http://www.bom.gov.au/weather-services/about/service-changes/vic_tfb_boundaries_changes2010.shtml; **e** tornado—A Oklahoma in 1999. (NOAA Photo Library via Creative Commons) http://www.google.com.au/imgres?imgurl=http://www.neontommy.com/sites/default/files/oklahoma%252520tornado.jpeg%3F1334350316&imgrefurl=http://www.neontommy.com/news/2012/04/tornadoes-storms-predicted-throughout-weekend&h=331&w=500&sz=103&tbnid=QZNBk6laYkDKGM:&tbnh=90&tbnw=136&prev=/search%3Fq%3Dcreative%2Bcommons%2Btornado%26tbm%3Disch%26tbo%3Du&zoom=1&q=creative+commons+tornado&usg=__TbHdZxDDBRIUWjFnofz1SlITZkI=&docid=NpTo2bg_KtFNoM&hl=en&sa=X&ei=eiWSUbCDHMSRkQWPooDIBg&sqi=2&ved=0CDgQ9QEwAw&dur=2215

species: Extinction risk from climate change includes marine or freshwater environments, explains the science behind the projections, sets forth new risk estimates for future climate change and considers the conservation and policy implications of the estimates. A review of the book in the British Journal of Entomology states "*it is no longer a question of whether or not we are in the midst of a mass extinction event, we are, the question now is, what can we do about it?*"

In an initial key paper Hansen et al. (1988) modeled the climate effects of rising GHG for the period 1958–1988. Scenario [A] assumes exponential growth in trace greenhouse gases; Scenario [B] assumes reduced linear growth of trace gases, and scenario [C] a rapid curtailment of trace gas emissions by the year 2000. In all three models peak temperatures reached the previous maximum Holocene levels.

8.2 The Great Carbon Oxidation Event

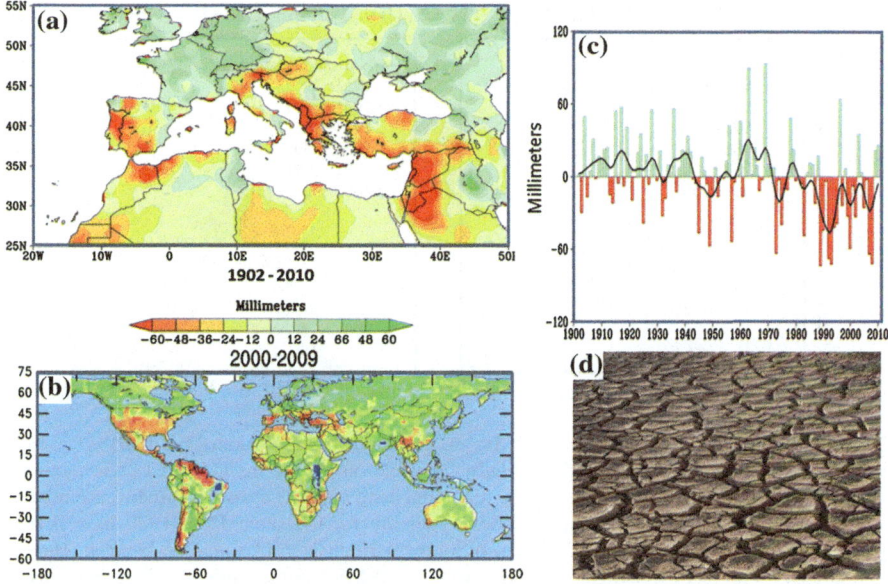

Fig. 8.14 a 2000–2009 droughts. The map uses a common measure, the Palmer Drought Severity Index, which assigns positive numbers when conditions are unusually wet for a particular region, and negative numbers when conditions are unusually dry. A reading of −4 or below is considered extreme drought. Regions that are *blue* or *green* will likely be at lower risk of drought, while those in the *red* and *purple* spectrum face more unusually extreme drought conditions. (https://www2.ucar.edu/atmosnews/news/2904/climate-change-drought-may-threaten-much-globewithin-decades). Courtesy Ivonne Mondragon, UCAR (University Corporation for Atmospheric Research; b Winter precipitation trends in the Mediterranean region for the period 1902–2010 c *Reds* and *oranges* highlight lands around the Mediterranean that experienced significantly drier winters during 1971–2010 than the comparison period of 1902–2010. b and c—Credit NOAA http://www.noaanews.noaa.gov/stories2011/20111027_drought.html; d drought (http://upload.wikimedia.org/wikipedia/commons/e/e1/Drought.jpg)

For model [A] the authors state "Our model results suggest that global greenhouse warming will soon rise above the level of natural climate variability". This observation was reported by James Hansen to the US Congress on 23 June, 1988. By 2012, allowing for the masking effect of sulphur aerosols, the unmasked temperature equivalent of the GHG rise has reached level remarkably close to that of Scenario [A] (Fig. 8.2). Hansen et al. (2008) regard a CO_2 level of ∼350 ppm as the maximum allowable before amplified feedbacks lead to climate tipping points beyond human control. Following the 1998 peak El Nino event (Fig. 8.18), the mean temperature rise rate declined relative to 1975–1998 warming rates due to a surge in SO_2 emissions (Fig. 8.6b) and a decline in sunspot activity. A dominance of La-Nina cycles during 2000–2012 (Fig. 8.18c) raises the question whether this phase is being enhanced by increasing ice melt driving the relatively cold Humboldt and California currents?

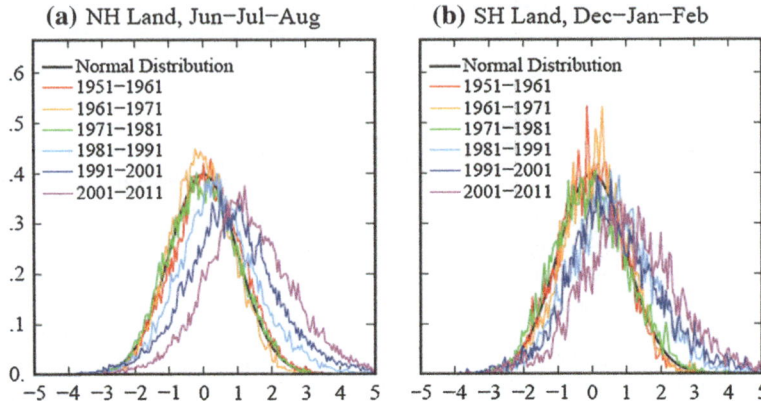

Fig. 8.15 Frequency of occurrence (y-axis) of local temperature anomalies divided by local standard deviation (x-axis) obtained by binning all local results for the indicated region and 11-year period into 0.05 frequency intervals. Area under each curve is unity. Anomalies are relative to 1951–1980 climatology, and standard deviations are for the same 1951–1980 base period. **a** Northern Hemisphere Land, Jun-Jul-Aug; **b** Southern Hemisphere Land, Dec-Jan-Feb (Hansen et al. 2012a, b Fig. 4; Proceedings National Academy of Science, vol. 109 no. 37, by permission (http://www.pnas.org/content/109/37/E2415/1.full)

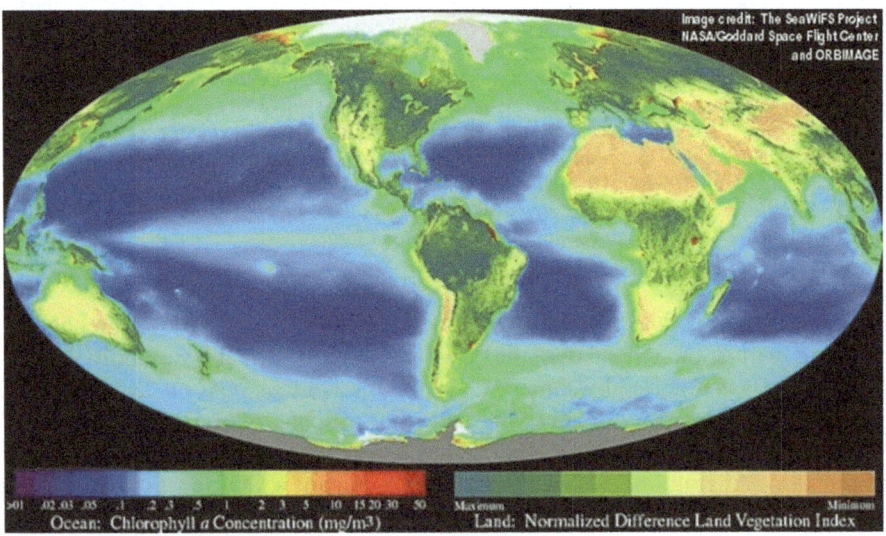

Fig. 8.16 Concentration of chlorophyll in the oceans (mg/m^3) representing the activity of phytoplankton (NASA http://www.esrl.noaa.gov/psd/). Note the enrichment of photosynthesis in cold water in polar and cold current-dominated oceanic regions. The density of vegetation on the continents is also shown

8.2 The Great Carbon Oxidation Event

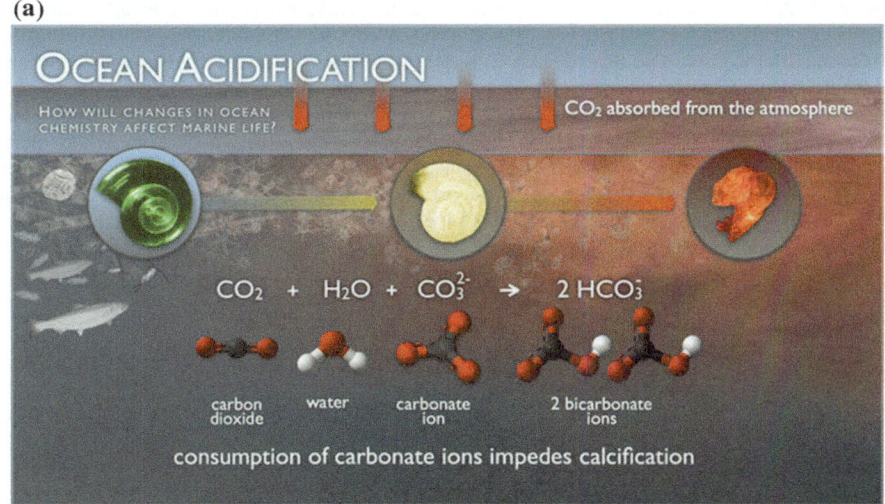

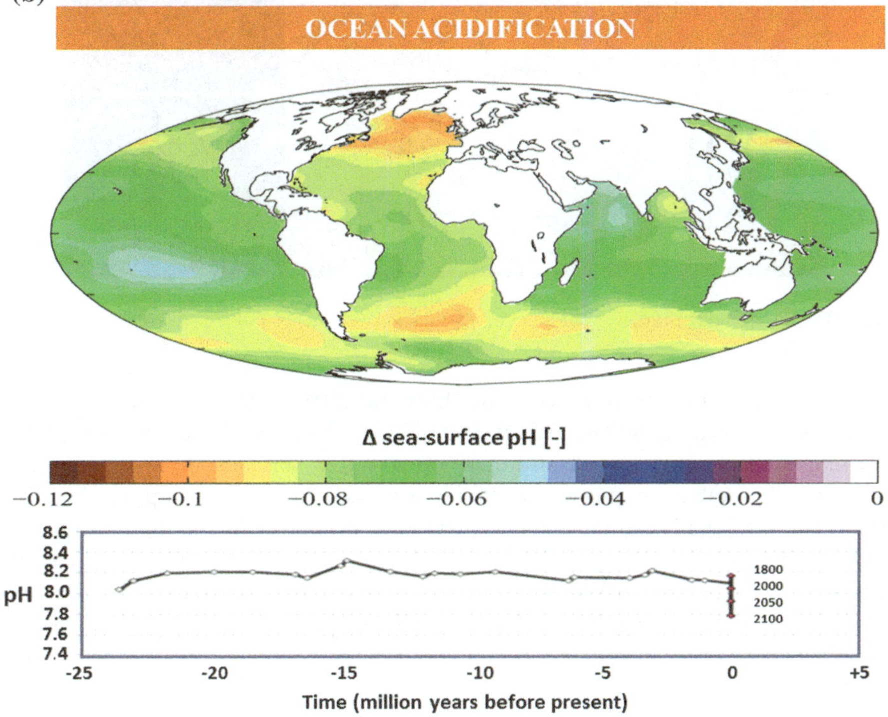

◀ **Fig. 8.17** A Chemical reaction of CO_2, water and carbonate ion producing bicarbonate ions, impeding calcification. http://www.pmel.noaa.gov/co2/file/Ocean+Acidification+Illustration **B** Ocean acidification. The map displays greater decline in pH in the cold sub-polar oceans (0.08–0.1 pH units) relative to tropical and subtropical waters (0.05–0.08 pH units), reflecting the stronger sequestration of atmospheric CO_2 in lower temperature waters (http://upload.wikimedia.org/wikipedia/commons/9/99/AYool_GLODAP_del_pH.png). **C** (**a**) The marine good chain http://www.arctic.noaa.gov/images/arctic_marine_food_web.jpg; **b** An oceanic phytoplankton bloom in the South Atlantic Ocean, off the coast of Argentina (NASA. http://en.wikipedia.org/wiki/File:Phytoplankton_SoAtlantic_20060215.jpg); **c** Pteropods—http://upload.wikimedia.org/wikipedia/commons/a/aa/Sea_butterfly.jpg

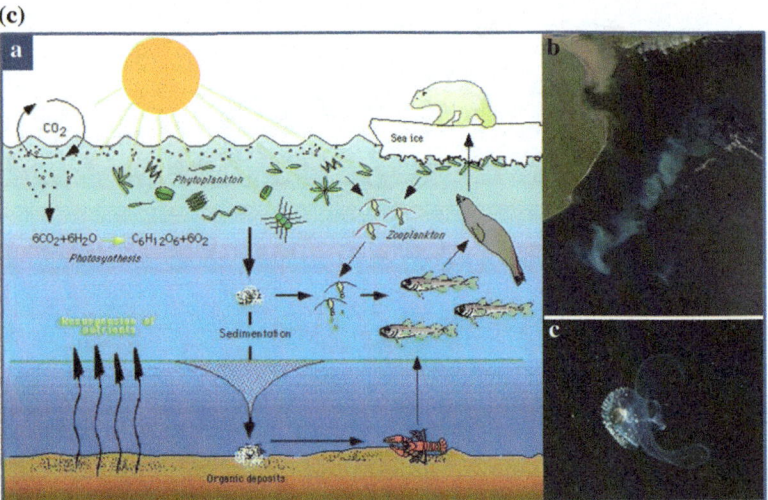

Fig. 8.17 (continued)

Peak temperatures were reached in 2006 and 2010 relative to the 1951–1980 respectively) (Fig. 8.18c) (NASA/GISS 2013). Positive feedbacks to global warming include summer exposure of open water surfaces in Polar Regions, replacing the reflective properties of ice surfaces infrared absorption by dark water. Hudson (2011) estimates the rise in radiative forcing due to total removal of Arctic summer sea ice as 0.7 Watt/m², which is close to the total of methane release since 1750 (~0.5 Watt/m²), although this amount would be in part offset by increased summer cloudiness. The subsequent increase in evaporation leads to the advance of cold vapor-laden fronts and to snow storms in the north Atlantic. Further positive feedbacks occur due to the release of methane from Arctic permafrost (total estimated as ~900 GtC), high-latitude peat lands (~400 GtC), Tropical peat lands (~100 GtC) and vegetation subject to fire and/or deforestation (CSIRO 2009) (Fig. 9.3).

By the onset of the third millennium *H. sapiens* has released more than 560 GtC from fossil biospheres and land clearing into the atmosphere, and is proceeding at

8.2 The Great Carbon Oxidation Event

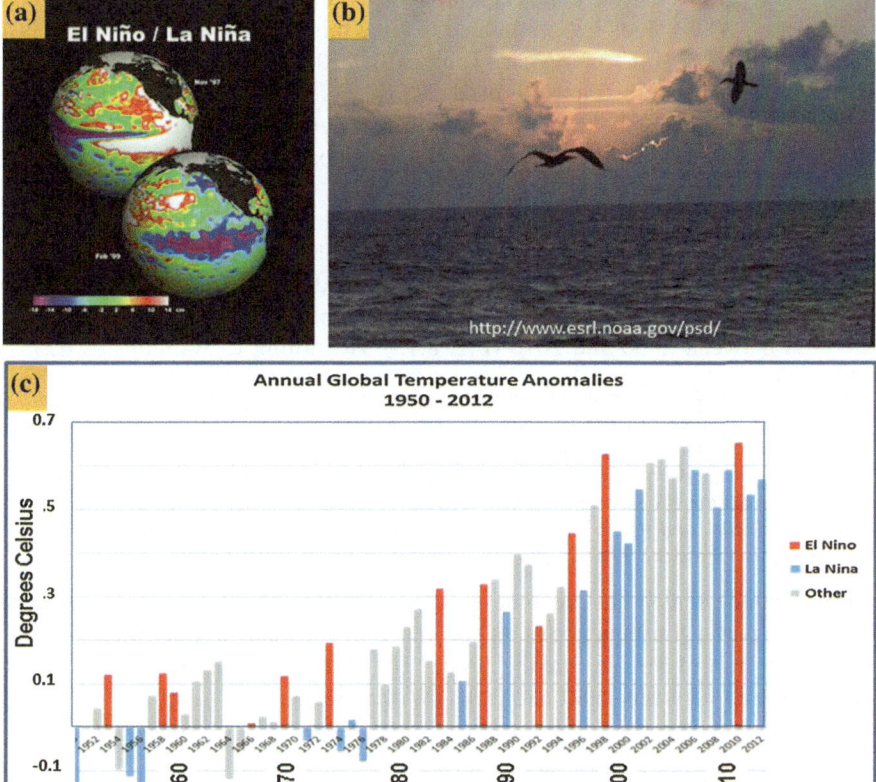

Fig. 8.18 Effects of global warming on the ENSO cycle. a The El-Nino phase (near-uniform temperatures and air-pressure along the equatorial Pacific) with consequent low or negative Southern Oscillation Index (SOI) and La-Nina phase (warm water driven westward with consequent high SOI) http://sealevel.jpl.nasa.gov/science/elninopdo/learnmoreninonina; b The Pacific Ocean http://www.esrl.noaa.gov/psd; c Annual global temperature anomalies as related to the ENSO cycle http://www.ncdc.noaa.gov/sotc/service/global/enso-global-temp-anom/201213.png; The El Niño phase of the cycle involves warmer-than-usual sea temperatures, great amounts of rainfall in South America, low atmospheric pressure, including marine life die-offs in the Pacific, hurricanes in Pacific islands and concurrent droughts in in India and Australia. La Nina conditions are driven by the cold Humboldt and California currents, rich in nutrients used by marine organisms, and involve dry conditions in the eastern Pacific. Rising global temperatures likely result in an increased intensity of the ENSO cycle, decline in cold currents and consequently a rise in the frequency of the El Nino phase

a rate of > 2 ppm CO_2 per year, unprecedented in geological history. Current economic reserves include 900 GtC Coal, 150 GtC Oil, 105 GtC Gas, and almost open ended reserves of tar sands oil, oil shale and coal-seam gas (Global Carbon Project 2012). A close parallel is the release to the atmosphere of some $\sim 2,000$

billion tons carbon at the Paleocene–Eocene boundary ~55 million years-ago (Zachos et al. 2008) (Fig. 3.3), estimated to have raised atmospheric CO_2 levels at a rate of about 0.11–0.13 ppm/year (Table A2), with attendant extinction of deep sea pelagic foraminifera (Panchuk et al. 2008). Other exceptions are global volcanic eruptions and asteroid impacts which have ignited regional to global wildfires (Durda and Kring 2004) and, in the case of impacts, excavated and vaporized carbon-rich sediments.

By the end of the 20th century and first decade of the 21st century climate change has ceased to be an abstract scientific notion and is manifesting itself through a spate of extreme weather events around the globe (Figs. 8.13, 8.14, 8.15). The number of reported disasters has increased from ~135 in 1980 to near−400 between 2000 and 2010 (Fig. 8.13). According to Williams (2012):

> *While the scientific debate spirals on into ever more intransigent spirals of obfuscation, the world continues to change around us. I've written before about the extraordinary number of extreme weather events last year, and the record number of temperature records set. 2011 is no different so far, with serious droughts developing in some parts of the world, and the US experiencing a vicious storm season … If it feels like there are far more hurricanes and floods these days, then your instincts are correct. The number of weather related disasters has increased dramatically in the last 30 years.*

The United States has become a focus of climate change-driven natural disasters, in part a consequence of its becoming an open corridor between the high-temperature Gulf of Mexico and the Arctic. The number of extreme weather events has quadrupled between 1980 and 2012 (Williams 2012).

8.3 The Sixth Mass Extinction of Species

In 2009 Joachim Hans Schellnhuber, Director of the Potsdam Climate Impacts Institute and Climate Advisor to the German Government, stated: *"We're simply talking about the very life support system of this planet"*. The consequences for the biosphere of accelerating climate change are discussed by Baronsky et al. (2013) in the following terms: *"Localized ecological systems are known to shift abruptly and irreversibly from one state to another when they are forced across critical thresholds. Here we review evidence that the global ecosystem as a whole can react in the same way and is approaching a planetary-scale critical transition as a result of human influence"* and *"Climates found at present on 10–48 % of the planet are projected to disappear within a century, and climates that contemporary organisms have never experienced are likely to cover 12–39 % of Earth. The mean global temperature by 2070 (or possibly a few decades earlier) will be higher than it has been since the human species evolved"*.

The decimation of habitats amounting to a crisis in the biosphere is manifest (Vitousek 1994; Casper 2009). A rapid polar-ward shift of climate zones is manifest. According to Xu et al. (2013) and NASA (2013) vegetation growth at Earth's northern latitudes increasingly resembles lusher latitudes to the south and

8.3 The Sixth Mass Extinction of Species

temperature at northern latitudes now resemble those found 4 degrees to 6 ° of latitude farther south as recently as 1982. Arctic sea ice and the duration of snow cover are diminishing, the growing season is getting longer and plants are growing to a greater extent. In the Arctic and boreal areas characteristics of the seasons are changing, leading to great disruptions for plants and related ecosystems. The Arctic's greenness is visible on the ground as an increasing abundance of tall shrubs and trees in locations all over the circumpolar Arctic. Greening in the adjacent boreal areas is more pronounced in Eurasia than in North America, driven by amplified greenhouse effects.

Plants and organisms display remarkable adaptation and regeneration powers where medium to long-term environmental changes occur. However, abrupt transitions and tipping points in the physical environment may exceed the adaptive capacity of some species, resulting in extinction. By 1950 near to 70 % of Mediterranean and temperate forests and near 50 % of tropical and sub-tropical forests were lost (Figs. 8.19a[b]). Projected changes in biodiversity need to discriminate between the effects of climate and land-use change effects (deChazal and Rounsevell 2009). Habitat loss leading to decreased species richness is the most common land-use change. Habitats most affected by climate change factors include grassland, shrubland, boreal forests, cool conifer forests and Tundra whereas habitats mostly affected by land clearing are tropical forests, warm mixed forests and temperate deciduous forests (Fig. 8.19a). Threatened and lost mammal species concomitant with loss of forest habitats, particularly pronounced in tropical and sub-tropical habitats, is documented in reports by the Millennium Ecosystem Assessment program (Fig. 8.19b).

In the oceans the marine food chain (Fig. 8.17c) may collapse in part under the pressures of decreasing pH and increasing temperatures. According to Veron (2008)

> if CO_2 levels are allowed to increase to 650–700 ppm, as is projected to occur later this century, a return to pre-industrial level will take a period of about 30,000 and 35,000 years. Initially acidification is buffered by bicarbonate–carbonate ion exchange, but once the buffers are overwhelmed the pH changes abruptly. The oceans will remain acidified until neutralized by the dissolving of marine carbonate rocks and the weathering of rocks on land, a hugely protracted process. When CO2 levels increase to 560 ppm, the Southern Ocean surface waters will be undersaturated with respect to aragonite, and the pH will be reduced by about 0.24 units—from almost 8.2 today to a little more than 7.9. At the present rate of acidification, all reef waters will have a $\Omega_{aragonite}$ of 3.5 or less by the middle of this century. Should CO2 levels reach 800 ppm later this century, the decrease will be 0.4 units and dissolved carbonate ion concentration will have decreased by almost 60 %. At that point all the reefs of the world will be eroding relics. The levels of CO_2 and pH predicted by the end of this century may not have occurred since the Middle Eocene, but the all-important rate of change we are currently experiencing has no known precedent. There can be no evolutionary solution for such a rate of change. Ultimately—and here we are looking at centuries rather than millennia—the ocean pH will drop to a point at which a host of other chemical changes, including anoxia, would be expected. If this happens, the state of the oceans at the end of K/T, or something like it, will become a reality and the Earth will enter the sixth mass extinction. Another 1–3 decades like our last will see the Earth committed to a trajectory from which there will be no escape.

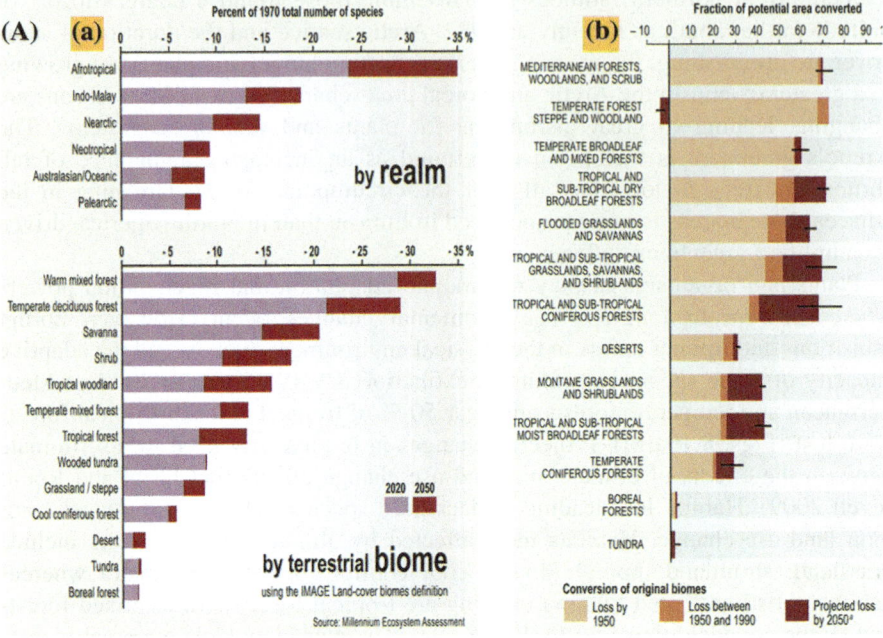

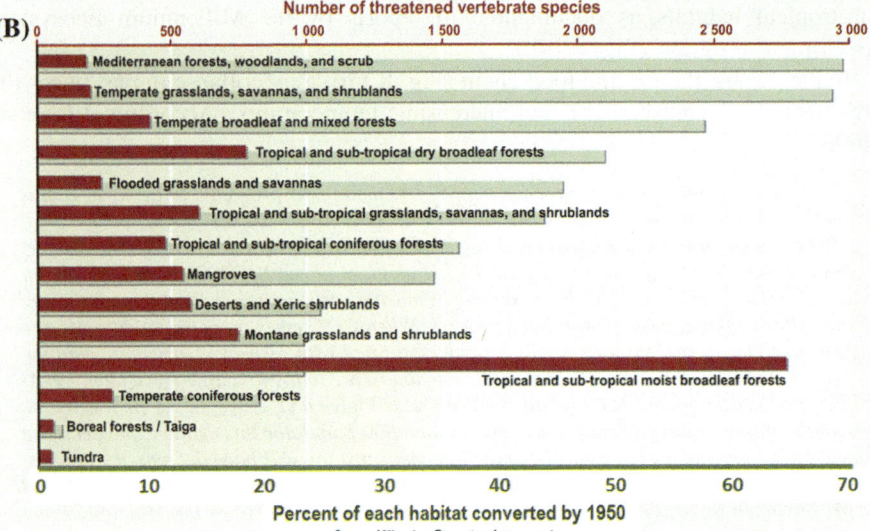

Fig. 8.19 A (a) Relative Loss of Biodiversity of Vascular Plants between 1970 and 2050 as a result of Land Use Change for Different Biomes (a biome is a major regional group of distinctive plant and animal communities) and Realms. Extinctions will occur between now and sometime after 2050, when populations reach equilibrium with remaining habitat. **b** Conversion of Terrestrial Biomes: It is not possible to estimate accurately the extent of different biomes prior to significant human impact, but it is possible to determine the "potential" area of biomes based on soil and climatic conditions. This Figure shows how much of that potential area is estimated to have been converted by 1950 (medium certainty), how much was converted between 1950 and 1990 (medium certainty), and how much would be converted under the four Millennium Assessment scenarios (low certainty) between 1990 and 2050. From "Ecosystems and human well-being—Biodiversity Synthesis". Millennium Ecosystems Assessment. http://www.unep.org/maweb/en/Synthesis.aspx. (World Resources Institute, by permission). **B** Threatened Vertebrates in 14 biomes ranked by the amount of their habitat converted by 1950. Percent of converted habitats—in *green*; Percent of threatened species—in *red*. From "Ecosystems and human well-being—Biodiversity Synthesis". Millennium Ecosystems Assessment. http://www.unep.org/maweb/en/Synthesis.aspx (World Resources Institute, by permission)

References

Baronsky et al (2013) Approaching a state shift in Earth's biosphere. Nature 486:52–58

Beerling DJ, Lomax BH, Royer DL, Upchurch GR, Kump LR (2002) An atmospheric pCO_2 reconstruction across the Cretaceous-tertiary boundary from leaf mega fossils. Proc Nat Acad Sci 99:7836–7840

Bowman DM et al (2009) Fire in the Earth system. Science 324:481–484

Broecker WC, Stocker TF (2006) The Holocene CO_2 rise: anthropogenic or natural? Eos 87:27–29

Casper K (2009) Changing ecosystems: effects of global warming. p 247.http://books.google.com.au/books?id=ZnUl4onKLs8C&printsec=frontcover&source=gbs_ge_summary_r&cad=0#v=onepage&q&f=false

Chandler DM et al (2013) Evolution of the subglacial drainage system beneath the Greenland Ice Sheet revealed by tracers. Nat Geosci 6:195–198

Crutzen PJ, Stoermer EF (2000) The 'Anthropocene'. Glob Change Newslett 41:17–18

CSIRO Report 2009/108. Permafrost melt poses major climate change threat http://www.csiro.au/Portals/Media/Permafrostclimate-change-threat.aspx

deChazal J, Rounsevella MDA (2009) Land-use and climate change within assessments of biodiversity change: a review. Glob Environ Change 19:306–315

Durda DD, Kring DA (2004) Ignition threshold for impact-generated fires. J Geophys Res 109:1–14

Gammage B (2012) The biggest estate on Earth: how Aborigines made Australia. Allen and Unwin, Crows Nest, p 384

Ganopolski A, Rahmstorf S (2002) Abrupt glacial climate changes due to stochastic resonance. Physics Rev Lett 88:3–6

Gibbon E (1788) The decline and fall of the Roman empire. http://books.google.com/books?id=f8-2ONV-foQC&pg=PA158

Global Carbon Project, Global Carbon Budget (2012) http://www.globalcarbonproject.org/carbonbudget/12/hl-full.htm

Hannah L (ed) (2011) Saving a million species: extinction risk from climate change. Island Press, Washington, D.C., p 432

Hansen J (2012) http://www.columbia.edu/~jeh1/mailings/2012/20120330_SlovenianPresident.pdf

Hansen JE (2012) http://www.ted.com/talks/james_hansen_why_i_must_speak_out_about_climate_change.html

Hansen J, Fung I, Lacis A, Rind R, Lebedeff S, Ruedy R, Russell G, Stone P (1988) Global climate changes as forecast by Goddard Institute for space studies 3D model. J Geophys Res 93:9341–9364

Hansen J, Sato M, Kharecha P, Lea DW, Siddall M (2007) Climate change and trace gases. Phil Trans Roy Soc 365A:1925–1954

Hansen J, Sato M, Kharecha P, Beerling D, Masson-Delmotte V, Pagani M, Raymo M, Royer DL, Zachos JC (2008) Target atmospheric CO_2: where should humanity aim? Open Atmos Sci J 2:217–231

Hansen J, Sato M, Kharecha P, von Schuckmann K (2012a) Earth's energy imbalance and implications. Atmos Chem Phys 11:13421–13449

Hansen J, Sato M, Ruedy R (2012b) Perception of climate change. Proc Nat Acad Sci www.pnas.org/cgi/doi/10.1073/pnas.1205276109 1-9

Hudson SR (2011) Estimating the global radiative impact of the sea-ice-albedo feedback in the Arctic. J Geophys Res 116:D16102

IPCC (2007) Contribution of working group I to the fourth assessment report of the Intergovernmental panel on climate change. http://www.ipcc.ch/publications_and_data/ar4/wg1/en/contents.html

Joos F, Gerber S, Prentice IC, Otto-Bliesner BL, Valdes PJ (2004) Transient simulations of Holocene atmospheric carbon dioxide and terrestrial carbon since the Last glacial maximum. Glob Biogeochem Cycles 18:GB2002

Jouzel J (2005) Stable carbon cycle—climate relationship during the late Pleistocene Sci 310:1313–1317

Kay CE (1994) Aboriginal overkill and native burning: implications for modern ecosystem management. W J Appl For 10:121–126

Kemp AC, Horton BP, Donnelly JP, Mann ME, Vermeere M, Rahmstorf S (2011) Climate related sea-level variations over the past two millennia. Proc Nat Acad Sci www.pnas.org/cgi/doi/10.1073/pnas.1015619108

Kutzbach JE, Ruddiman WF, Vavrus SJ, Philippon G (2010) Climate model simulation of anthropogenic influence on greenhouse-induced climate change (early agriculture to modern): the role of ocean feedbacks. Climatic Change 99:351–381

Le Quere C (2012) The global carbon budget 1959–2011. Earth Syst Sci Data Discuss 5:1107–1157

Lenton TM, Held H, Kriegler E, Hall JW, Lucht W, Rahmstorf S, Schellnhuber HJ (2009) Tipping points in the Earth system. Proc Nat Acad Sci 105:1786–1793

Lewis HT (1985) Why Indians burned: specific versus general reasons. In: Lotan JE et al (ed) Proceedings–symposium and workshop on wilderness fire, Nov 15–18, 1983, Missoula, Montana

McCulloch MT (2008) Current state and future of the oceans and marine life in a high CO_2 world. In: Imagining the real: life on a greenhouse Earth. Manning Clark House conference, Manning Clark House, Canberra, p 25–27

Munich Re (2012) Cat report underscores extreme weather in U.S.; Sandy Losses http://www.insurancejournal.com/news/international/2013/01/03/275865.htm

Murphy DM, Solomon S, Portmann RW, Rosenlof KH, Forster PM, Wong T (2009) An observationally based energy balance for the Earth since 1950. J Geophys Res 114:D17107. doi:10.1029/2009JD012105

NASA (2013) Amplified Greenhouse effect shifts North's growing seasons. http://www.nasa.gov/home/hqnews/2013/mar/HQ_13-069_Northern_Growing_Seasons.html

NASA/GISS (2013) Temperatures. http://data.giss.nasa.gov/gistemp/

NOAA (2013) Mouna Loa CO_2. http://www.esrl.noaa.gov/gmd/ccgg/trends/

Panchuk K, Ridgwell A, Kump LR (2008) Sedimentary response to Paleocene-Eocene thermal maximum carbon release: a model-data comparison". Geology 36:315–318

PRISM (2012) Global warming analysis, pliocene research, interpretation and synoptic mapping. US Geol. Surv. http://geology.er.usgs.gov/eespteam/prism/index.html

Pyne SJ (1982) Fire in America: A cultural history of wild and rural fire. Princeton University Press, Princeton

Pyne SJ (1995) World fire: the culture of fire on Earth. Henry Holt and Company, New York

Rahmstorf S (2007) A semi-empirical approach to projecting future sea level rise. Science 315:368–370

Rignot E, Velicogna I, Van der Broeke MR, Monaghan A Lenaerts J (2011)Acceleration of the contribution of the Greenland and Antarctic ice sheets to sea level rise. Geophys Res Lett 38:L05503

Ruddiman WF (2003) Orbital insolation, ice volume, and greenhouse gases. Quatern Sci Rev 22:1597–1629

Russell EWB (1983) Indian-set fires in the forests of the Northeast United States. Ecology 64:78–88

Sheuyangea A, Oba G, Weladji RB (2005) Effects of anthropogenic fire history on savanna vegetation in northeastern Namibia. J Environ Manag 75:189–198

Siddall M, Rohling EJ, Almogi-Labin A, Hemleben CH, Meischner D, Schmelze I, Smeed DA (2003) Sea-level fluctuations during the last glacial cycle. Nature 423:853–858

Smith SJ, van Aardenn J, Klimont Z, Andres RJ, Volke A, Delgado Arias S (2011) Anthropogenic sulfur dioxide emissions: 1850–2005. Atmos Chem Phys 11:1101–1116

Steffen W, Crutzen PJ, McNeill JR (2007) The anthropocene: are humans now overwhelming the great forces of nature? Ambio 36:614–621

Stocker B, Strassmann K, Joos F (2010) Sensitivity of Holocene atmospheric CO_2 and the modern carbon budget to early human land use: analysis with a process-based model. Biogeosciences Discuss 7:921–952

Thomas CD (2004) Extinction risk from climate change. Nature 427:145–148

Velicogna I (2009) Increasing rates of ice mass loss from the Greenland and Antarctic ice sheets revealed by GRACE. Geophys Res Lett 36(19):L19503

Veron C (2008) Mass extinctions and ocean acidification: biological constraints on geological dilemmas. Coral Reefs 27:459–472

Vitousek PM (1994) Beyond global warming: ecology and global change. Ecology 75:1861–1876

Williams J (2012) The number of natural disasters is on the rise. http://makewealthhistory.org/2011/05/30/the-number-of-natural-disasters-is-on-the-rise/

Xu L et al (2013) Temperature and vegetation seasonality diminishment over northern lands. Nat Clim Change 3(4):1–6

Zachos J, Pagani M, Sloan L, Thomas E, Billups K (2001) Trends, rhythms, and aberrations in global climate 65 Ma to present. Science 292:686–693

Zachos J, Dickens GR, Zeebe RE (2008) An early Cenozoic perspective on greenhouse warming and carbon-cycle dynamics. Nature 451:279–283

Chapter 9
An Uncharted Climate Territory

Abstract Since the onset of the industrial age in the 18th century and accelerating since the mid 1980s, the release of more than 560 billion ton of carbon (GtC) through industrial emission and land clearing has triggered unprecedented developments in the terrestrial climate at a rate faster by an order of magnitude than natural geological warming events. Whereas comparisons can be made with the Paleocene-Eocene Thermal Maximum of ~55 Ma, the scale and rate of modern global warming may compare more closely with those triggered by major volcanic and asteroid impact events. The non-linear nature of current climate change, multiple feedbacks and their synergy are driving the climate to uncharted territory and possible tipping points.

By 2012–2013 the atmospheric CO_2 concentration, measured at Mauna Loa, has risen by 2.89 ppm (394.45–397.34 ppm (NOAA 2013), implying a return to Pliocene-like atmospheric radiative forcing at a rate unprecedented in the Cainozoic geological record, except for instant rises of GHG associated with asteroid impacts. A continuation of carbon emission at rates of near-30 $GtCO_2$/year or higher implies a rise of atmospheric CO_2 to levels of about 600 ppm and higher (Figs. 9.1a, b, 9.2)—unrecorded since the Eocene pre-32 Ma-ago (Zachos et al. 2001) (Fig. 3.1a, b; Fig. 3.2). Amplifying feedbacks from vulnerable carbon pools (Figs. 9.3 and 9.4) are capable of raising atmospheric CO_2 to levels which resulted in mass extinction at 55 Ma-ago (Zachos et al. 2008) (Fig. 3.3).

A perspective on such development is allowed by the mid-Miocene (~16 Ma) warm period. Only small ice caps existed during this period while sea levels were about 40–50 m higher than at present. However, whereas the transition from end-Miocene conditions to the present took place over some 5.2 million years, the rise of mean global temperature since the 18th century would hardly allow many species to adapt. Whereas species may adapt to shifting states of the climate, the extreme rate at which radiative greenhouse gas forcing is rising—~3 ppm during 2012–2013—would not allow many species the time to adapt.

The extent to which global warming and ocean acidification will develop depends on a number of factors, including:

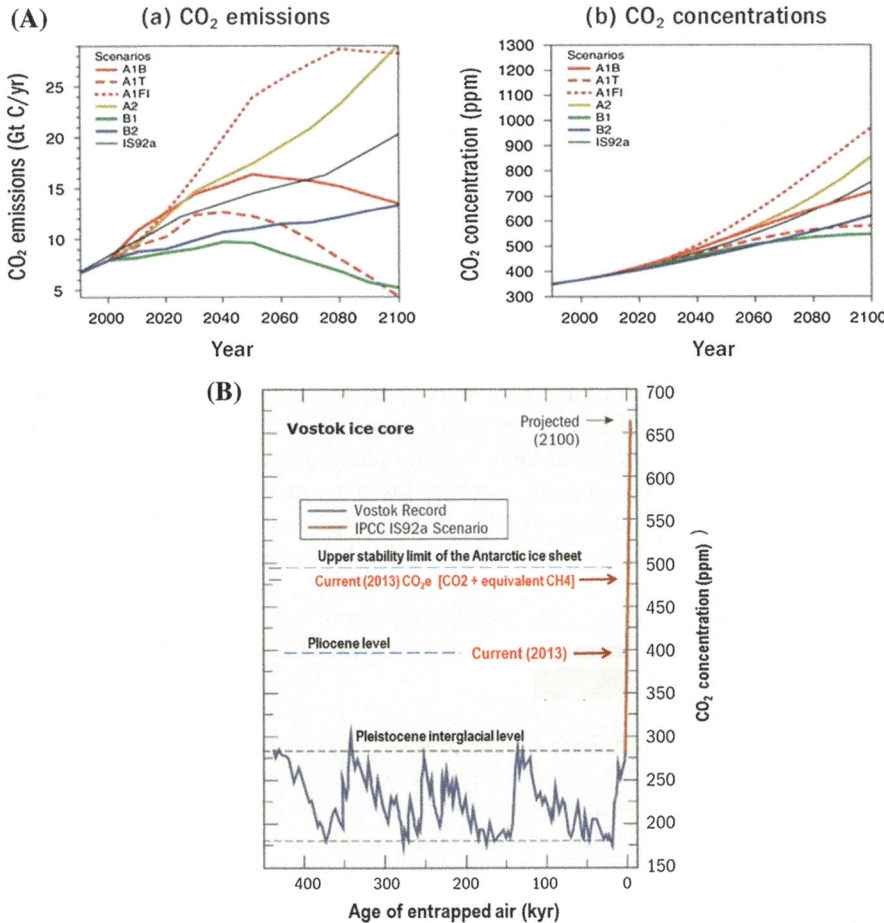

Fig. 9.1 A (**a**) CO_2 emissions and (**b**) atmospheric CO_2 concentrations (Global Carbon Project, Canadell et al. 2013, Fig. 16 http://www.globalcarbonproject.org/science/index.htm). **B** The Vostok ice core record for atmospheric concentration (Petit et al. 1999) and the 'business as usual' prediction used in the IPCC third assessment (Climate change 2001). The current and Pliocene CO_2 levels, the CO_2-e (total CO_2^+ equivalent CH_4) and the stability threshold of the Antarctic ice sheet are indicated. Global Carbon Project, Canadell et al. 2013, Fig. 2 http://www.globalcarbonproject.org/science/index.htm. Courtesy P. Canadell)

1. Reserves of fossil fuels large enough exist for atmospheric CO_2 levels to reach more than 600 ppm (Fig. 9.1a, b) or, for exploitation of non-conventional resources, more than 1,000 ppm (Fig. 9.2). The extent to which the estimated resources of coal (>10,000 GtC), oil (~700 GtC) and gas (>2,000 GtC) would be combusted and released to the atmosphere within the next few decades and centuries depends critically on global economic developments and the effects of extreme weather events on industry and transport systems. A central looming

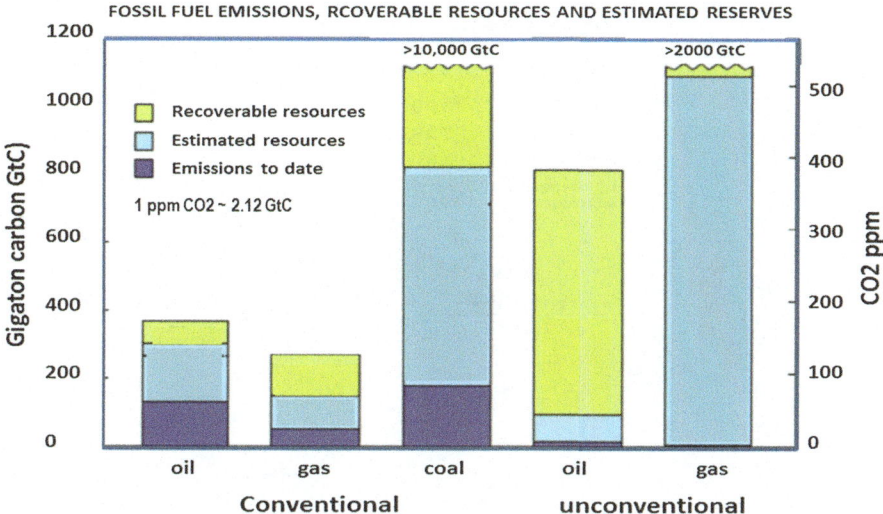

Fig. 9.2 Estimates of fossil fuel resources and equivalent atmospheric CO_2 levels, including (1) emissions to date; (2) estimated reserves, and (3) recoverable resources (2.12 GtC = ~1 ppm CO_2). (Hansen 2012a; Hansen et al. 2012b, Fig. 1; http://www.columbia.edu/~jeh1/mailings/2012/20120127_CowardsPart1.pdf

factor arises from increasing droughts in otherwise agriculturally productive parts of the world, including southern Europe, northern India and Australia (Figs. 8.14, 9.6a). Rising temperatures through the 21st century (Fig. 9.6a) can only enhance this trend, leading to global droughts (Figs. 8.14, 9.6a) and food shortages. With this perspective carbon emissions are inherently self-limiting, leading to an inevitable decline of their sources. The extent to which amplifying feedbacks from current and future CO_2 levels would contribute to further rises in atmospheric greenhouse gases, global warming and the crossing of climate tipping points is difficult to quantify.

2. Vast amount of carbon stored in the arctic and boreal regions, estimated at over 1.3 trillion tons of frozen carbon (Fig. 9.3) is more than double that previously estimated (Canadell 2009) (Fig. 9.3). Carbon in permafrost is found largely in northern regions including Canada, Greenland, Mongolia, Russia, Scandinavia, and USA. Radioactive ^{14}C carbon dating shows that most of the carbon dioxide currently emitted by thawing soils in Alaska is released from carbon frozen thousands of years-ago and decomposes when soils thaw under warmer conditions. The evidence to date shows that carbon in permafrost is likely to play a significant role in the 21st century climate given the large carbon deposits, the readiness of organic matter to release greenhouse gases when thawed, and the fact that high latitudes will experience the largest relative increase in air temperature of all regions (Canadell 2009). Oxidation of ~6,000 GtC from fossil fuel, vulnerable vegetation and methane pools in permafrost and peat land would consume about 2.10^{13} tons O_2, i.e., about 1 % of the atmospheric

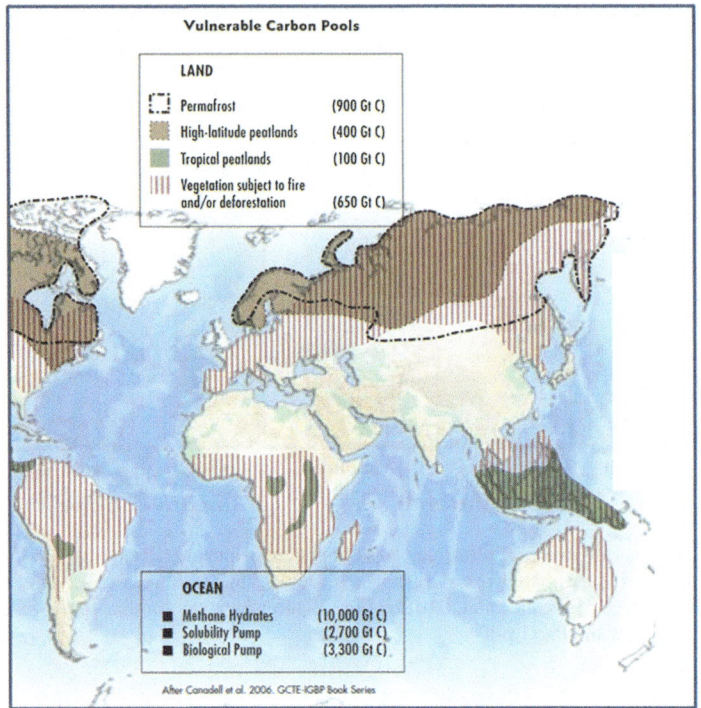

Fig. 9.3 Vulnerable carbon sinks. (1) Land: Permafrost—600 GtC, High-latitude peatlands—400 GtC, tropical peatlands—100 GtC, vegetation subject to fire and/or deforestation—650 GtC, (2) Oceans: Methane hydrates—10,000 GtC; Solubility pump—2,700 GtC; Biological pump—3,300 GtC (After Canadell et al. 2006 GCTE-IGBP Book series; The Global carbon cycle; UNESCO-SCOPE policy briefs; Vol 2. Courtesy P. Canadell)

inventory of $1.4.10^{15}$ ton O_2. Given oxygen atmospheric residence time of 4,500 years and an oxygen cycle in the order of 3.10^{14} ton O_2 per-year for the atmosphere and biosphere, long term decline in photosynthesis would result in several percent decline in atmospheric O_2, similar to developments during geological greenhouse states (Berner et al. 2007). The direct dependence of photosynthesis (Fig. 8.16), ocean temperature, pH (Fig. 8.17) and O_2 levels render the current trend (Fig. 8.17b) deleterious to the marine food chain (Fig. 8.17c).

3. The dissociation and rate of release of methane deposits from permafrost and frozen Arctic lakes and sediments as a feedback to global warming is critical to near-future greenhouse gas trajectories. The magnitude of methane deposits (Fig. 9.3), gas reservoirs (Fig. 9.4) and coal seam gas reserves and their release, either as feedbacks to global warming or due to human exploitation, constitutes key factors threatening the future of the present state of the biosphere.

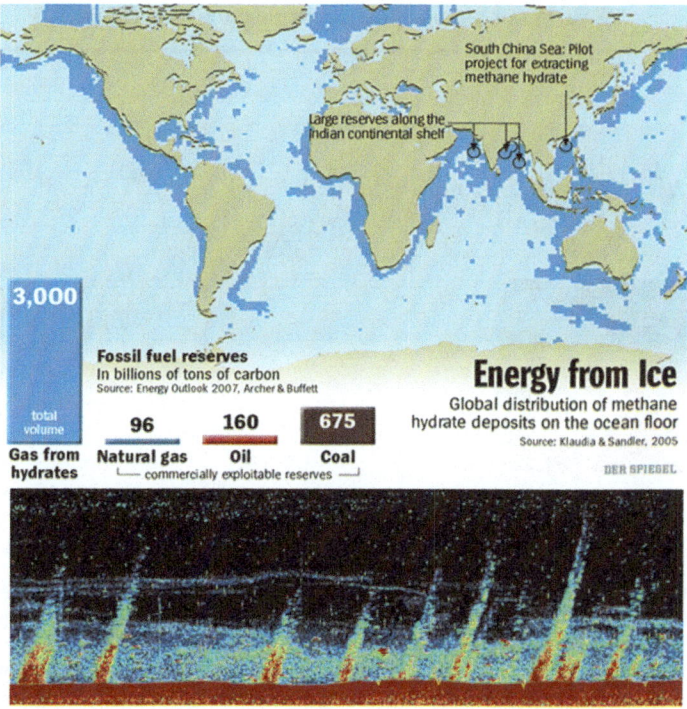

Fig. 9.4 a Global distribution of methane hydrate deposits on the ocean floor and commercially exploitable carbon reserves (Naval Research Laboratory, by permission; http://www.nrl.navy.mil/research/nrl-review/2002/chem-biochem-research/coffin/), **b** Image of methane bubbling from the sea floor http://news.bbc.co.uk/2/hi/8205864.stm http://www.nature.com/scitable/knowledge/library/methane-hydrates-and-contemporary-climate-change-24314790. Westbrook et al. (2009) Courtesy Graham Westbrook

4. The North Atlantic Thermohaline Current (NATH) overturning and deep water formation has been declining since about 1950 (Fig. 9.5). Estimates vary between 1 and 4 Sverdrup (1 Sv = $10^6 m^3.s^{-1}$).[1] Advanced melting of Greenland ice (Fig. 8.10) and increasing fluxes of cold ice-melt water into the North Atlantic would lead to further weakening and eventual collapse of the NATH. The timing and synergy of tipping points around the world, including the effects of feedbacks from fire on atmospheric CO_2, outlined by Lenton et al. 2009, include melting of Arctic Sea, Greenland and west Antarctic ice, permafrost and Tundra loss, changes in the ENSO, changes in West African and Indian monsoon, Boreal forest and Amazon dieback, (Fig. 9.6d).
5. Changes in the ENSO reverse the trend recorded from the Pliocene to the Pleistocene glacial-interglacial cycles (Fig. 3.6b), signifying a return to climate

---

[1] 1 Sverdrup = one million cubic meters of water flow per second.

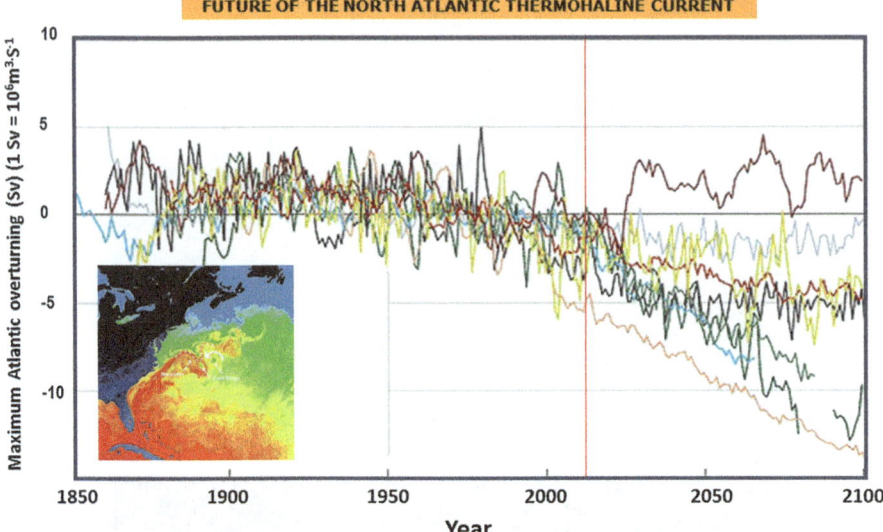

Fig. 9.5 a Observed and simulated change in ocean circulation strength with time performed by several research institutes using different coupled atmosphere–ocean models and common forcing scenarios (Stute et al. 2001. Proceedings National Academy of Science, by permission http://www.pnas.org/content/98/19/10529.full.pdf+html); inset - satellite infrared image of the North Atlantic thermohaline current (NASA—http://www.treehugger.com/renewable-energy/swarms-of-floating-marine-turbines-could-harvest-power-from-the-gulf-stream.html

conditions of about 3 million years-ago, including drought conditions in the southwest Pacific, storminess in northwest South America and gradual greening of parts of the Sahara.

The consequences of a synergy between the processes outlined above, underpinned by the unprecedented rates at which these processes are taking place (Fig. 8.9), can hardly be quantified. Whereas the precise timing of global climate tipping points remains difficult to define, the current intensification of extreme weather events around the globe, expressed by the increased variability of climate events (Rahmstorf and Coumou 2011; Hansen et al. 2012a, b) (Figs. 8.13, 8.15) may express tipping elements in the climate. A rise of $> 0.4°$ C in mean sea surface temperature during 1960–2010 (NASA/GISS 2013) (Fig. 8.8Ba) has led to enhanced intensity and in some instances frequency of hydrological evaporation/precipitation cycle in several parts of the world, ensuing in intensification of cyclones and floods (Fig. 8.13). The rise in land temperatures results in increased frequency and intensity of heat waves and fires, at a rate about two to threefold during 1980–2008 (Munich Re-Insurance 2012) (Fig. 8.13).

As stated by the IPCC (2012): "*Models project substantial warming in temperature extremes by the end of the 21st century. It is virtually certain that increases in the frequency and magnitude of warm daily temperature extremes and*

9 An Uncharted Climate Territory

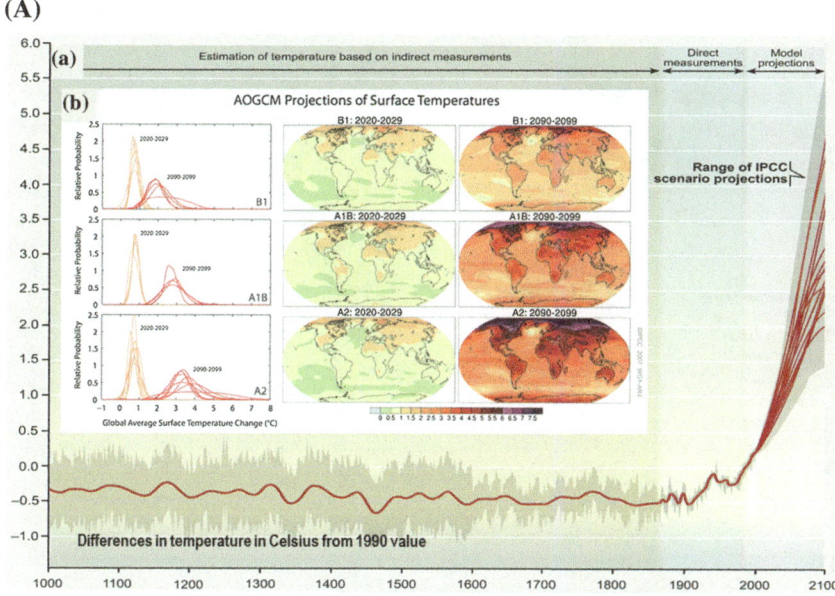

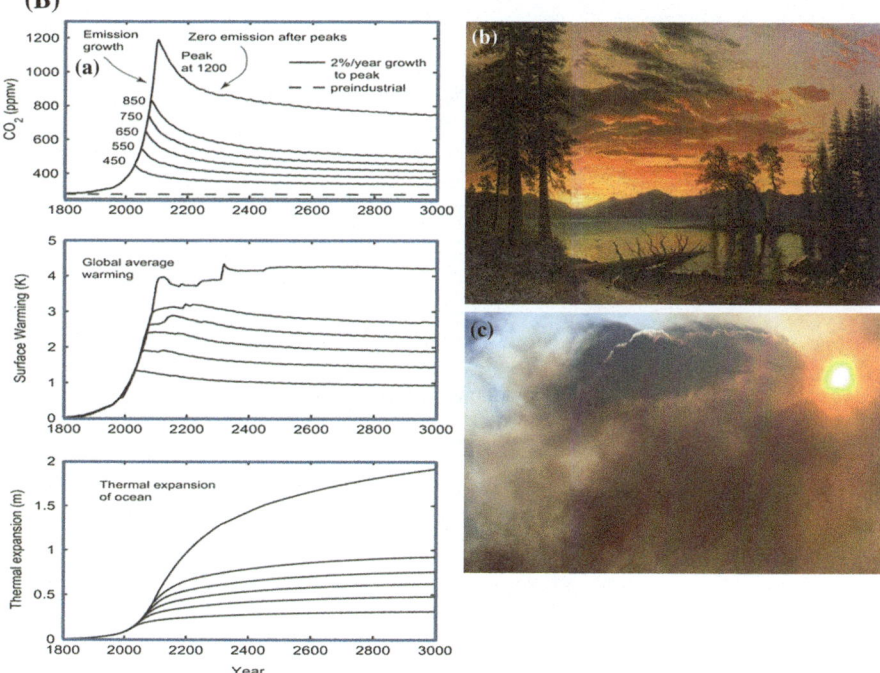

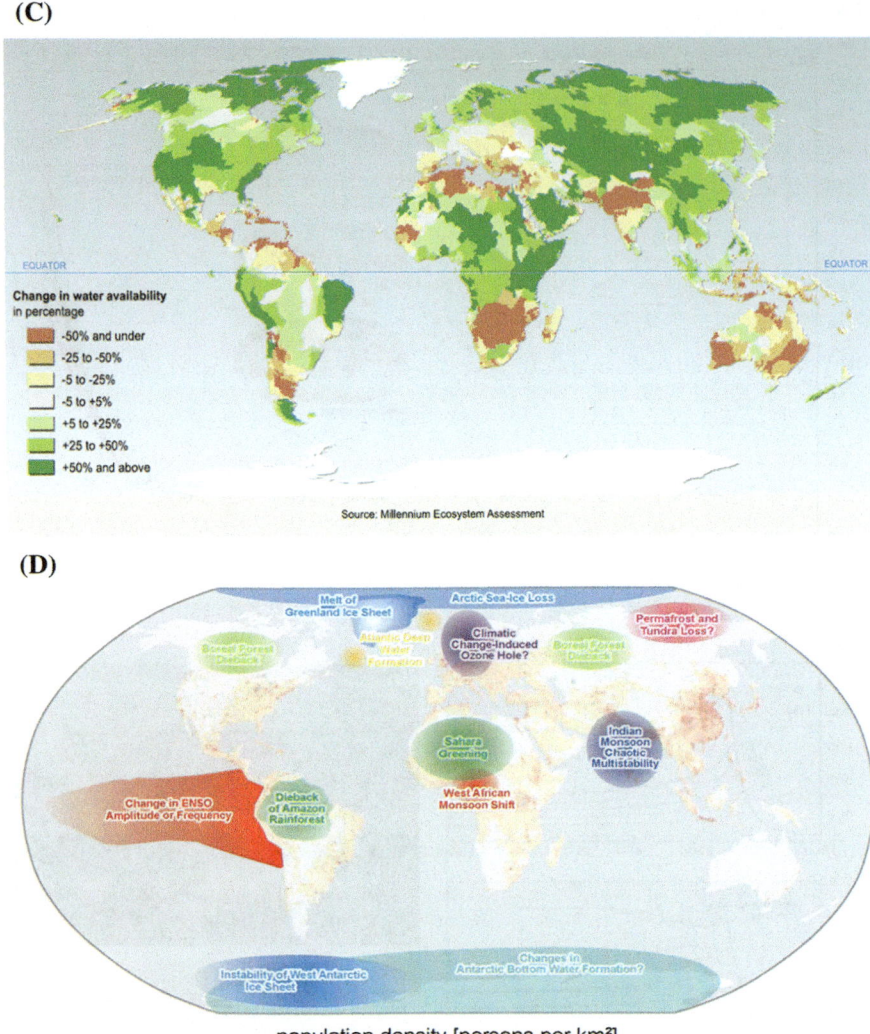

Fig. 9.6 (continued)

◀ **Fig. 9.6 A (a)** Historical and Projected Variations in Earth's Surface Temperature. Estimated global temperature averages for the past 1,000 years, with projections to 2100 depending on various plausible scenarios for future human behaviour (http://www.unep.org/maweb/en/Synthesis.aspx. World Resources Institute, by permission); (http://www.unep.org/newscentre/default.aspx?DocumentID=2697&ArticleID=9293&l=en); **b** Atmosphere–Ocean Global Circulation Model (AOGCM) projections of surface temperatures. Projected surface temperature changes for the early and late 21st century relative to the period 1980–1999. The central and right panels show the Atmosphere–Ocean General Circulation multi-Model average projections for the B1 (*top*), A1B (*middle*) and A2 (*bottom*) SRES scenarios averaged over decades 2020–2029 (*centre*) and 2090–2099 (*right*). (IPCC AR4 2007. The Physical Science Basis. Working group I contribution to the fourth assessment report of the intergovernmental panel on climate change. Cambridge University Press. Figure SPM-6. by permission) http://www.ipcc.ch/publications_and_data/ar4/wg1/en/spmsspm-projections-of.html) **B (a)** Carbon dioxide and global mean climate system changes relative to preindustrial conditions in 1765 from illustrative model, the Bern 2.5CC EMIC, whose results are comparable to the suite of assessed EMICs. Climate system responses are shown for a ramp of CO_2 emissions at a rate of 2 %/year to peak CO_2 values of 450, 550, 650, 750, 850, and 1,200 ppmv, followed by zero emissions. Results have been smoothed using an 11 year running mean (Solomon et al. 2009), **b** Sunset—by Rama Devi—Nature's Contemplative Notes http://www.wikinut.com/img/1fuf13v.wj9wadbz/Albert-Bierstadt-Public-domain-Wikimedia-Commons, **c** Wildfire, Saratoga Springs, June 22, 2012 (photo: Melissa Hincha-Ownby, flickr creative commons) http://sittingwithsorrow.typepad.com/.a/6a0154363a7e57970c01774360f7ad970d-pi. **C** 2009–2030 droughts. The map uses a common measure, the Palmer Drought Severity Index, which assigns positive numbers when conditions are unusually wet for a particular region, and negative numbers when conditions are unusually dry. A reading of −4 or below is considered extreme drought. Regions that are blue or green will likely be at lower risk of drought, while those in the red and purple spectrum face more unusually extreme drought conditions (Courtesy Ivonne Mondragon, UCAR (University Corporation for Atmospheric Research https://www2.ucar.edu/atmosnews/news/2904/climate-change-drought-may-threaten-much-globe-within-decades) **D** Map of potential policy-relevant tipping elements in the climate system, overlain on global population density. Subsystems indicated could exhibit threshold-type behaviour in response to anthropogenic climate forcing, where a small perturbation at a critical point qualitatively alters the future fate of the system. They could be triggered this century and would undergo a qualitative change within this millennium. Excluded from the map are systems where threshold appears inaccessible this century, e.g., East Antarctic Ice Sheet, or the qualitative change would appear beyond this millennium, e.g., marine methane hydrates. Question marks indicate systems whose status as tipping elements is particularly uncertain (Lenton et al. 2009. Proceedings National Academy of Science USA, by permission)

decreases in cold extremes will occur in the 21st century on the global scale. It is very likely that the length, frequency and/or intensity of warm spells, or heat waves, will increase over most land areas. Based on the A1B and A2 emissions scenarios, a 1-in-20 year hottest day is likely to become a 1-in-2 year event by the end of the 21st century in most regions, except in the high latitudes of the Northern Hemisphere, where it is likely to become a 1-in-5 year event". Rahmstorf and Coumou (2011) conducted a statistical analysis of the relations between long-term climate trends and the incidence of extreme weather events, finding that the number of record-breaking heat events increases approximately in proportion to the ratio of warming trend to short-term standard deviation, or variability. Short-term variability decreases the number of heat extremes, whereas a longer term climatic warming trend increases it.

Since about 2006 the rate of atmospheric methane rise increased to more than 7 ppb/year (Fig. 8.3). By 2010 reports of advanced release of methane from the East Siberian Arctic Shelf (ESAS) have been communicated by Shakova et al. (2010). The summary report reads: "*Remobilization to the atmosphere of only a small fraction of the methane held in East Siberian Arctic Shelf (ESAS) sediments could trigger abrupt climate warming, yet it is believed that sub-sea permafrost acts as a lid to keep this shallow methane reservoir in place. Here, we show that more than 5000 at-sea observations of dissolved methane demonstrates that greater than 80 % of ESAS bottom waters and greater than 50 % of surface waters are supersaturated with methane regarding to the atmosphere. The current atmospheric venting flux, which is composed of a diffusive component and a gradual ebullition component, is on par with previous estimates of methane venting from the entire World Ocean. Leakage of methane through shallow ESAS waters needs to be considered in interactions between the biogeosphere and a warming Arctic climate.*" In an interview (Shakova 2010) clarifies that the concentration of atmospheric methane over parts of the East Siberian shelf is 8–10 % higher than global concentration, the highest recorded in the Pleistocene ice cores, and that the amount of methane released from the East Siberian shelf is greater than the total released from the world oceans.

Projections of 21st century climate trends suggest that, excepting the masking effects by sulphur aerosols, the climate is fast-tracking toward conditions such as existed during the Pliocene (~ 5.3–2.6 Ma) and the peak Miocene (~ 16 Ma), when temperatures were ~ 2 °C to ~ 4 °C higher than pre-industrial Holocene levels. A development of pre-Oligocene (pre-34 Ma) ice-free conditions, when atmospheric CO_2 levels exceeded $\sim 500 \pm 50$ ppm (Fig. 3.2), depends on the growth rate of carbon emission and atmospheric CO_2. According to Hansen et al. (2012a, b) "*Burning all fossil fuels would create a different planet than the one that humanity knows. The palaeoclimate record and ongoing climate change make it clear that the climate system would be pushed beyond tipping points, setting in motion irreversible changes, including ice sheet disintegration with a continually adjusting shoreline, extermination of a substantial fraction of species on the planet, and increasingly devastating regional climate extremes*". However, it is unlikely that the world's fossil fuel reserves, estimated as >13,000 GtC (Hansen 2012a, b) (Fig. 9.2), could be combusted as deteriorating global climate, extreme weather events and their consequences for agriculture and industry can only reduce the use of fossil fuels, rendering emissions self-limiting.

Should the global community choose to focus its remaining resources on mitigation of carbon emissions, climate geo-engineering efforts (Royal Society 2009) and adaptation to the fast deteriorating climate, some of the methods outlined in Table D1 may apply. A species which has placed a man on the moon should also be able to develop a range of effective greenhouse gas sequestration methods in order to protect its home planet. Since the onset of sulphur emissions from coal and oil, masking the effects of greenhouse gas radiative forcing by about -1.1 to -1.5 Watt/m_2 since 1750 (IPCC 2007; Murphy et al. 2009; Hansen et al. 2012a, b), near-to 50 % of global warming has been transiently mitigated, constituting an unintended geo-engineering measure. Regional variations occur, with large concentrations of aerosols over industrial centers in the Northern Hemisphere. With the acceleration of climate change since the 1980s an increase in sulphur emissions is being considered in order to cool the planet, along with a range of other mitigation and CO_2 draw-down methods (Table D1).

Principal observations include (Glikson 2012):

1. Stratospheric sulfur injections are both short-lived and destructive in terms of ocean acidification and retardation of the monsoon and of precipitation over large parts of the Earth, including Africa, southern and Southeast Asia.
2. Retardation of solar radiation through space sunshades is of limited residence time and would not prevent ocean acidification from ongoing carbon emission.
3. The dissemination of ocean iron filings and temperature exchange through pipe systems is likely ineffective in transporting CO_2 to safe water depths.
4. CO_2 sequestration using soil carbon, biochar and possible chemical methods such as sodium trees, combined with rapid decline in industrial CO_2 emissions, can in principle help slow down, and in future even reverse, the current rise in atmospheric CO_2 toward mean global temperatures above 2 degrees Celsius.
5. Budgets on a scale of military spending (>$20 trillion since WWII) are required in order to attempt to retard current trend toward likely tipping points, including increasing release of methane from permafrost and Arctic sediments.
6. Top priority ought to be given to fast track testing of soil carbon burial/biochar methods, chemical ('sodium trees') and incentives for invention of new CO_2 sequestration methods.

Should a global effort at mitigation and adaptation not occur, rapid warming and ocean acidification would affect the base of the food chain (Fig. 8.17) and thereby survival of large mammals and many other species. Warm acid oceans would be severely depleted in corals, phytoplankton, krill and higher marine life. As occurred during the K-T extinction small burrowing mammals could fare better. Many bird species, deprived of many of their migratory routes and nesting homes, would suffer demise. A yet higher scale of Mass Extinction would ensue from a nuclear coup-de-grace, increasingly likely on a planet stressed by environmental crises, compounding a 'greenhouse summer' with a 'nuclear winter'. The enhanced hydrological cycle, favoring extensive growth in expanded tropical and subtropical regions, could lead to the opening of new ecological niches and proliferation of resistant life forms such as insect species.

Ultimate consequences of global warming are alluded to in Ward (2007) 'Under A Green Sky', portraying three scenarios:

1. Drastic cut in carbon emissions in an attempt to keep atmospheric CO_2 below 450 ppm;
2. Atmospheric CO_2 level reaches 700 ppm by the year 2100 while sea level rises on the scale of meters and the North Atlantic conveyer collapses;

3. by 2100 CO_2 levels reach 1,100 ppm and temperatures 10 °C above pre-industrial level, ensue in total melt of polar and glacier ice and in deep ocean anoxia, culminating in the sixth great mass extinction, as has already commenced (Figs. 9.7a, b).

The longevity and potential irreversibility of atmospheric CO_2 modeled by Solomon et al. (2009) and Eby et al. (2009) imply long term tropical conditions (Fig. 9.6b) postdating a cessation of emissions by at least 1,000 years. Following cessation of emissions the rate of temperature transfer to the oceans declines and atmospheric temperatures decline at a slower rate. A rise of atmospheric CO_2 to peak levels of 450–600 ppm would result in dry season reduction in rainfall and continental dust-bowl conditions in several regions (Solomon et al. 2009). Multi-millennial simulations by Eby et al. (2009) find the period required to absorb anthropogenic CO_2 strongly depends on the total amount of emissions and that for emissions similar to known fossil fuel reserves, the time to absorb 50 % of the CO2 is more than 2,000 years. The long-term climate response appears to be independent of the rate at which CO_2 is emitted over the next few centuries. The lifetime of the surface air temperature anomaly might be as much as 60 % longer than the lifetime of anthropogenic CO_2 and two-thirds of the maximum temperature anomaly will persist for longer than 10,000 years, suggesting the consequences of anthropogenic CO_2 emissions will persist for many millennia (Eby et al. 2009).

According to Berger and Loutre (2002) current climate change is leading to an exceptionally long interglacial period ahead (Fig. 9.8). These authors state: *'The present day CO_2 concentration (397 ppm by 2012) is already well above typical interglacial values of ~ 290 ppmv. This study models increases to up to 750 ppmv over the next 200 years, returning to natural levels by 1,000 years. The results suggest that, under very small insolation variations, there is a threshold value of CO_2 above which the Greenland ice sheet disappears. The climate system may take 50,000 years to assimilate the impacts of human activities during the early third millennium. In this case an 'irreversible greenhouse effect could become the most likely future climate. If the Greenland and west Antarctic ice sheets disappear completely, then today's 'Anthropocene' may only be a transition between the Quaternary and the next geological period'.*

Flooding of low-lying river systems (Ganges, Mekong, Yellow River, Nile, Indus, Tigris/Euphrates, Mississippi, Rhine, Thames), would decimate major agrarian populations and flooding of tropical river valleys (e.g., Amazon and Congo) destroy vast tracts of rainforest. Depending how far global warming proceeds and the extent to which humanity's nuclear arsenal is released, by accident or design, some members of the species may survive, mostly those genetically adapted to extreme conditions in remote parts of the globe. Rising Arctic temperatures will still leave Polar Regions within human body temperature comfort zones. There, small human clans such as the Inuit may survive, provided ocean acidification does not completely destroy the base of the food chain. The intensified hydrological cycle will enhance precipitation in the Siberian Taiga,

9 An Uncharted Climate Territory

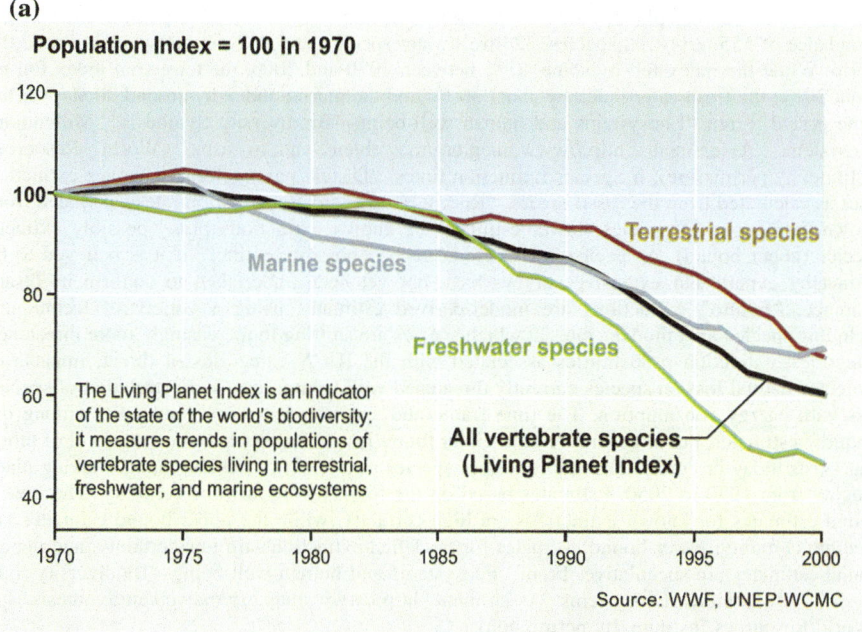

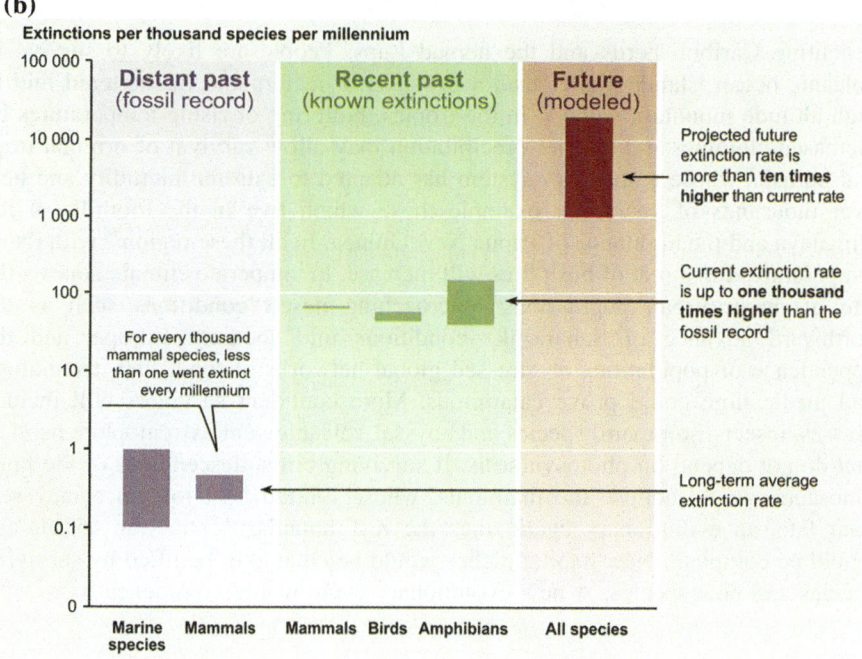

◀**Fig. 9.7 a** The living planet index 1970–2000. The index currently incorporates data on the abundance of 555 terrestrial species, 323 freshwater species, and 267 marine species around the world. While the index fell by some 40 % between 1970 and 2000, the terrestrial index fell by about 30 %, the freshwater index by about 50 %, and the marine index by around 30 % over the same period. From "Ecosystems and human well-being—Biodiversity Synthesis". Millennium Ecosystems Assessment http://www.unep.org/maweb/en/Synthesis.aspx (World Resources Institute, by permission); **b** Species Extinction Rates. "Distant past" refers to average extinction rates as calculated from the fossil record. "Recent past" refers to extinction rates calculated from known extinctions of species (lower estimate) or known extinctions plus "possibly extinct" species (upper bound). A species is considered to be "possibly extinct" if it is believed to be extinct by experts but extensive surveys have not yet been undertaken to confirm its disappearance. "Future" extinctions are model-derived estimates using a variety of techniques, including species-area models, rates at which species are shifting to increasingly more threatened categories, extinction probabilities associated with the IUCN categories of threat, impacts of projected habitat loss on species currently threatened with habitat loss, and correlation of species loss with energy consumption. The time frame and species groups involved differ among the "future" estimates, but in general refer to either future loss of species based on the level of threat that exists today or current and future loss of species as a result of habitat changes taking place roughly from 1970 to 2050. Estimates based on the fossil record are low certainty. The lower-bound estimates for known extinctions are high certainty, while the upper-bound estimates are medium certainty; lower-bound estimates for modelled extinctions are low certainty, and upper-bound estimates are speculative. From "Ecosystems and human well-being—Biodiversity Synthesis". Millennium Ecosystems Assessment http://www.unep.org/maweb/en/Synthesis.aspx (World Resources Institute, by permission)

benefiting Caribou herds and the nomad Laps. People are likely to survive in volcanic ocean islands higher than a few tens of meters and in sheltered mid to high altitude mountain valleys. In the tropics, buffering of rising temperatures by increased clouding and heavier precipitation may allow survival of original tropical humans whose respiratory system has adapted to extreme humidity and heat over thousands of years, for example those which live in the foothills of the Himalaya and the highlands of Papua New Guinea. In all these regions, with rising temperatures the threat of bush fires will increase. In temperate climate zones—the site of major urban populations—encroaching desert conditions such as the northward advance of Sahara-like conditions into southern Europe, and the dependence of populations on stressed global networks of food, fuel, technology and medication, could prove calamitous. More confident survivors will include grasses, insects, some bird species and abyssal volcanic vent extremophile habitats that do not depend on photosynthesis. If surviving birds, descendants of the fated dinosaurs, would eclipse the mammals, whose limited heat tolerance may seal their fate, an evolutionary cycle since the K-T boundary extinction 65 Ma-ago would be complete. New habitat niches would be created to be filled by surviving species and new species. A new evolutionary cycle would commence.

9 An Uncharted Climate Territory

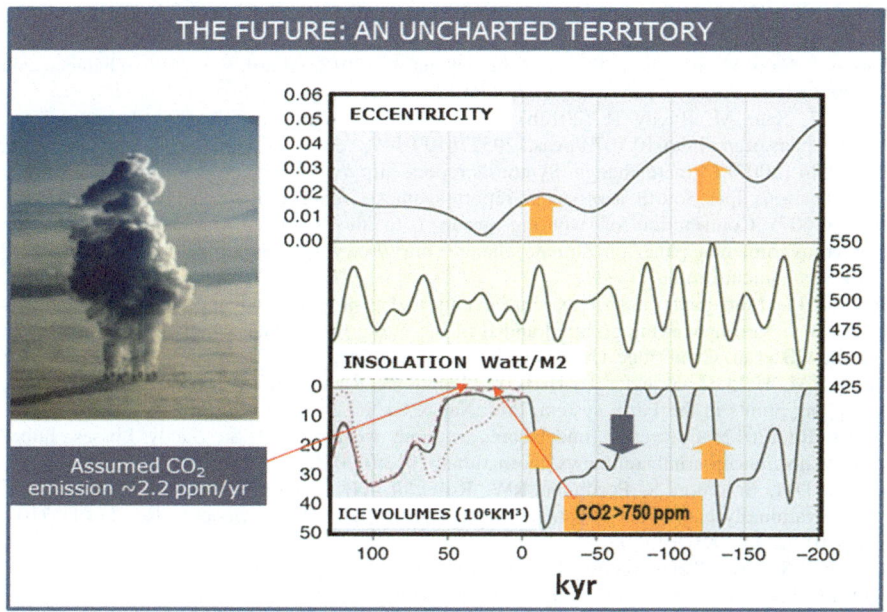

Fig. 9.8 An exceptionally long interglacial ahead? (Berger and Loutre 2002). Long-term variations of eccentricity (*top*), June insolation at 65°N (*middle*), and simulated Northern Hemisphere ice volume (*increasing downward*) (*bottom*) for 200,000 years before the present to 130,000 from now. Time is negative in the past and positive in the future. For the future, three CO_2 scenarios were used: last glacial-interglacial values (*solid line*), a human-induced concentration of 750 ppmv (*dashed line*), and a constant concentration of 210 ppmv (*dotted line*)

References

Berger A, Loutre MF (2002) An exceptionally long interglacial ahead? Science 297:1287–1288
Berner RA, Vanderbrook JM, Ward PD (2007) Oxygen and evolution. Science 316:557–558
Canadell et al (2006) GCTE-IGBP Book series: the global carbon cycle. UNESCO-SCOPE policy briefs. Vol 2
Canadell J et al (2013) Global carbon project. http://www.globalcarbonproject.org/science/index.htm
Canadell P (2009) Super-size deposits of frozen carbon threat to climate change. http://www.eurekalert.org/pub_releases/2009-06/gcp-sdo_1063009.php
Climate change (2001) Synthesis report. http://www.ipcc.ch/pdf/climate-changes-2001/synthesis-spm/synthesis-spm-en.pdf
Eby N, Zickfeld K, Montenegro A, Archer D (2009) Lifetime of anthropogenic climate change: millennial time scales of potential CO2 and surface temperature perturbations. J Climate 22:2501–2511
Glikson AY (2012) Geoengineering the climate? A southern hemisphere perspective: a Symposium, Australian academy of science, national committee for earth system science, p 42–43
Hansen J (2012a) http://www.columbia.edu/~jeh1/mailings/2012/20120330_SlovenianPresident.pdf

Hansen JE (2012b) http://www.ted.com/talks/james_hansen_why_i_must_speak_out_about_climate_change.html

Hansen J, Sato M, Kharecha P, von Schuckmann K (2012a) Earth's energy imbalance and implications. Atmos Chem Phys 11:13421–13449

Hansen J, Sato M, Ruedy R (2012b) Perception of climate change. Proc Nat Acad Sci www.pnas.org/cgi/doi/10.1073/pnas.1205276109 1–9

IPCC AR4 (2007) Climate change. Synthesis report. http://www.ipcc.ch/publications_and_data/publications_ipcc_fourth_assessment_report_synthesis_report.htm

IPCC (2007) Contribution of working group i to the fourth assessment report of the intergovernmental panel on climate change. http://www.ipcc.ch/publications_and_data/ar4/wg1/en/contents.html

IPCC (2012) Managing risks of extreme events and disasters to advance climate change—a special report of working groups I and II of the intergovernmental panel on climate change. Field CB et al. Cambridge University Press, Cambridge, p 582

Lenton TM, Held H, Kriegler E, Hall JW, Lucht W, Rahmstorf S, Schellnhuber HJ (2009) Tipping points in the Earth system. Proc Nat Acad Sci 105:1786–1793

Munich Re (2012) Cat report underscores extreme weather in U.S.; Sandy Losses. http://www.insurancejournal.com/news/international/2013/01/03/275865.htm

Murphy DM, Solomon S, Portmann RW, Rosenlof KH, Forster PM, Wong T (2009) An observationally based energy balance for the Earth since 1950. J Geophys Res 114:D17107. doi:10.1029/2009JD012105

NASA/GISS (2013) Temperatures. http://data.giss.nasa.gov/gistemp/

NOAA (2013) Mouna Loa CO2. http://www.esrl.noaa.gov/gmd/ccgg/trends/

Petit JR et al (1999) 420,000 years of climate and atmospheric history revealed by the Vostok deep Antarctic ice core. Nature 399:429–436

Rahmstorf SR, Coumou D (2011) Increase of extreme events in a warming world. Proc Nat Acad Sci 108:17905–17909

Royal Society (2009) Geo-engineering the climate: science, governance and Uncertainty ISBN 978-0-85403-773-5

Shakova N (2010) http://hot-topic.co.nz/siberian-seabed-methane-first-numbers/

Shakova N et al (2010) Extensive methane venting to the atmosphere from sediments of the east siberian arctic shelf. Science 327:1246–1250

Solomon S, Plattner GK, Knutti R, Friedlingstein P (2009) Irreversible climate change due to carbon dioxide emissions. Proc Nat Acad Sci 0812721106 28 Jan 2009

Stute M, Clement A, Lohmann G (2001) Global climate models: past, present, and future. PNAS Proc Nat Acad Sci 98:10529–10530

Ward PD (2007) Under a green sky: global warming, the mass extinctions of the past, and what 514 they can tell us about our future. Harper Collins, New York, p 242

Westbrook GK et al (2009) Escape of methane gas from the seabed along the West Spitsbergen continental margin. Geophys Res Lett 36:L15608

Zachos J, Pagani M, Sloan L, Thomas E, Billups K (2001) Trends, rhythms, and aberrations in global climate 65 Ma to present. Science 292:686–693

Zachos J, Dickens GR, Zeebe RE (2008) An early cenozoic perspective on greenhouse warming and carbon-cycle dynamics. Nature 451:279–283

Chapter 10
Homo Prometheus

Abstract A species able to magnify its entropy effect in nature by orders of magnitude, as the genus Homo has done through mastery of fire and the splitting of the atom, needs to be a perfectly wise and controlled species, lest its invention gets out of hand.

> *For we are the local embodiment of a Cosmos grown to self-awareness. We have begun to contemplate our origins: star-stuff pondering the stars; organised assemblages of ten billion-billion-billion atoms considering the evolution of atoms; tracing the long journey by which, here at least, consciousness arose. Our loyalties are to the species and the planet.* **WE** *speak for the Earth. Our obligation to survive is owed not just to ourselves but also to that Cosmos, ancient and vast, from which we spring.*

(Carl Sagan, Cosmos, 1980).

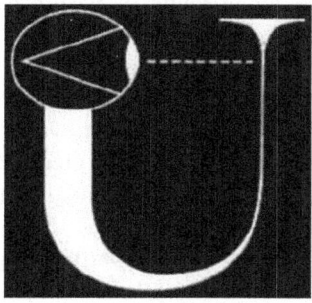

For a biological species to magnify its effect on entropy within the biosphere by orders of magnitude and to develop cerebral power that allows it to become the intelligent eyes through which the Universe explores itself, as represented by the U figure by John Archibald Wheeler, where a human eye explores beginnings of the Universe, hints at yet unknown natural laws. In terms of such laws the phenomenon of life, manifested all the way from self-repairing DNA-RNA biomolecules to an intelligence allowing insights into the Universe, remains unexplained. When Carbon, hydrogen, oxygen and nitrogen combined into the four DNA species, adenine, cytosine, guanine and thymine (de Silva and Williams 1991), a four billion years-long journey was embarked on, from microbes to the human brain, on a planetary surface subject to repeated volcanic eruptions and extraterrestrial bombardment. It has not been foreknown, until recently, that human intelligence, achieved in the process, will itself give rise to one of the most severe mass extinction events in planetary history. According to Wallace Broecker (1987) *The inhabitants of planet Earth are quietly conducting a gigantic experiment. We play Russian roulette with climate and no one knows what lies in the active chamber of the gun.*

Only recently have some members of the species Homo sapiens come to realize that, just as the individual life process depends on the oxygen-carbon cycle, via its lungs, so does the biosphere depend on the oxygen and carbon cycles of the atmosphere. Having mastered fire, Homo sapiens began to recruit elements from the periodic table as building blocks for its cultural evolution, including iron, gold and uranium. The discovery of smelting provided the species with tin and lead (> 6,500 BP), copper and bronze (> 5,500 BP), iron (> 3,000 BP) and other metals, allowing production of new tools, ploughs for sowing excess grain, silver and gold for the crafting of ornaments, symbolizing wealth and facilitating trade. Swords and spears elevated homicide to new levels. Subsequent iron-based machines allowed the industrial revolution and vast-scale industrial killings—called "war".

Gold, a soft low-melting (1,064 °C) volatile metal which symbolizes the excess resources required for buying power, feeding slaves or sending men to war, manifests the mythological aspiration of the human mind. Prior to the rise of mammon only limited quantities of loot could be carried across land and sea but once precious metals assumed token value of wealth and thereby power, they became motivators for wars, destroying people and civilizations, as became the fate of the Aztecs and the Inca of the Americas. A space visitor would watch with disbelief how gold is mined from 3 km deep beneath the Witwatersrand only to be buried in underground safes at Fort Knox. The traveler would be amazed how communities have been hijacked by the agents of greed and chaos represented by the share market, the common denominator of human conduct.

Societies ruled by tribal shamans, feudal war lords, speculators, thugs and habitual liars have opened Pandora's Box, perpetrating mass murders called "war", poisoning natural habitats for the mythological sake of ultimate power and the after-life. Since the mastery of fire, culminating in the fission of uranium ^{235}U, plutonium (^{239}Pu—half-life 24,360 years) accumulates in the biosphere while the carbon–oxygen balance of the atmosphere, lungs of the Earth, continues to be destroyed. In Albert Einstein's words *The splitting of the atom changed everything, save man's mode of thinking; thus we drift towards unparalleled catastrophe.*

Homo sapiens has been experimenting with the fate of the Earth. According to Albert Speer, German physicists, apprising the Nazi government of the possible development of an atom bomb in the spring of 1942, noted a reservation by Werner Heisenberg regarding a potential conflagration engulfing the atmosphere. The same awesome possibility, fusion of atmospheric nitrogen and oceanic hydrogen, turning the planet into a chain-reacting bomb, was considered by Edward Teller, Robert Oppenheimer, Arthur Compton, Hans Bethe and other physicists in connection with the H-bomb. Whereas calculations indicated an atmospheric conflagration was unlikely, a finite possibility remained, but the Bikini H-bomb tests went ahead. In the wake nuclear trials on the inhabitants of two Japanese cities, with lateral proliferation the possibility of global nuclear conflagration becomes a probability and a near-certainty in an environmentally stressed world. The Damocles sword of Mutual Assured Destruction (MAD) strategy remains. The hapless inhabitants of planet Earth have been given a non-choice between accelerating global heating and a nuclear winter. Experiments with the fate of Earth continue, and although the Hadron Collider has been deemed safe (Hadron Collider 2008), the consequences of further science fiction-like experiments dreamt by ethic-free people remain unknown.

Once the climate stabilized in the Holocene, allowing production of excess food, a fearful human mind has been unleashed, drafting slaves to construct pyramids for immortality and soldiers for murderous orgies to appease the gods. Echoing infanticide by rival warlord baboons, the butchering of children on Aztec altars and the sacrifice of generations in endless wars signifies a Saturn-like species devouring its sons. The Tower of Babel story tells of humans confused by language that can tell both truths and falsehoods. Nature is full of examples of creatures seducing their prey and destroying their host. Prehistoric and pantheistic humans revered animals, even while hunting and sacrificing them, whereas monotheism looks to heaven and space for salvation, leaving nature as but a corridor to higher realms. Having lost a sense of reverence toward Earth, there is no evidence humans are about to rise above the realm of perceptions, dreams, myths, legends and denial (Koestler 1986).

Planetcide emerges from dark recesses of the prehistoric mind, from the fears of humans watching the flames around camp fires, yearning for immortality. The transformation of tribal warriors into button-pushing automatons remains shrouded in mystery. Perhaps it is too much to expect any living species to possess the wisdom and responsibility required to control its own inventions. Charity remains, mostly among the wretched of the Earth, where empathy is learnt through suffering. From the Romans to the Third Reich, the rapacious brutality of the legions surpasses that of small marauding tribes bound by traditions. Under false flags empires never cease to kill peasants in their rice fields and palm groves. Unintended anthropogenic global heating threatens to turn the entire planet into an open-air inferno on the strength of a Faustian Bargain.

A child of Orwellian *Newspeak,* planetcide challenges every social system, faith, ideal or philosophy humans ever held. As in the *War of the Worlds* by H.G. Wells, free-thinking cells are destroyed by the parent organism. Ideals of democracy are invariably undermined by the rise of plutocracy, subterranean drug rings and weapon smuggling networks, while the minds of children are poisoned by commercial and political propaganda machines. Ultimately the rise of relativistic moral philosophies, justifying the unjustifiable, means that, without ethics, Homo sapiens cannot survive.

What would be the geological legacy of technological civilizations millions of years hence? Erosion, fires and subsequently the next glacial period, albeit delayed, would grind the stone and cement structures of cities into sand and clay. A telltale marker horizon of corroded metals, including gold and plutonium, precious stones and glass will form in the oceans. Already radioactive anomalies have been detected in clay beds and ice varves in several parts of the world (Grossman 2011).

By the end of the first decade of the 2nd Millennium it was becoming clear H. sapiens was not going to undertake a meaningful attempt to slow-down, arrest or reverse global warming consequent upon the discharge to the atmosphere of some 560 billion ton of carbon from the Earth crust. The bulk of its extra-resources continue to be poured into the military, entertainment, sports, gambling, electronic gaming and drugs poisoning the young's mind. With a majority oblivious to the fast changing climate, misinformed by vested interests and their media outlets, betrayed by cowardly leaders and discouraged by the sheer magnitude of the event, beyond human power, only a few scientists and nature lovers remain, becoming the subject of witch hunts, while humanity is drifting into unparalleled catastrophes.

Existentialist philosophy allows perspectives into, and ways of coping with, what defies rational contemplation. If looking into the sun may result in blindness, according to little-understood poetic aphorism, the human insights into nature entail a terrible price. Ethical and cultural assumptions of free will, which may operate to an extent in individual lives, do not govern the behavior of societies or nations, let alone a species. Despite all of the above, in an existentialist sense hope is essential for survival on the individual scale. Going through the dark night of the soul, individuals may experience the emergence of a conscious dignity devoid of illusions, grateful for a fleeting moment of awareness. *Having pushed a boulder up the mountain all day, turning toward the setting sun, we must consider Sisyphus happy* (Albert Camus 1942; The Myth of Sisyphus).

References

Broecker W (1987) Unpleasant surprises in the greenhouse?. Nature 328:123–126
Camus A (1942) The myth of sisyphus. Alfred A Knopf, New York, p 120
de Silva JRF, Williams RJP (1991) The biological chemistry of the elements. Clarendon Press, Oxford
Grossman E (2011) Radioactivity in the ocean: diluted, but far from harmless. Environment 360. http://e360.yale.edu/feature/radioactivity_in_the_ocean_diluted_but_far_from_harmless/2391/
Hadron Collider (2008) LHC switch-on fears are completely unfounded. The Institute of Physics. PR 48(08). 5 Sept 2008
Koestler A (1986) Janus: a summing up. Picador Books, London

Epilogue
The 'Life Force'

"My suspicion is that the Universe is not only queerer than we suppose, but queerer than we can suppose" J. B. S. Haldane.

Earth constitutes a unique planet in the Solar system and most likely far beyond. Drake and Dava (1992) estimate the frequency of technical civilizations in the Milky Way galaxy in terms of

$$N = R * fp * ne * f\ell * fi * fc * L.$$

N = the number of civilizations in our galaxy with which communication might be possible (i.e., which are on our current light cone); R = the average rate of star formation per year in our galaxy; fp = the fraction of those stars that have planets; ne = the average number of planets that can potentially support life per star that has planets; $f\ell$ = the fraction of the above that actually go on to develop life at some point; fi = the fraction of the above that actually go on to develop intelligent life; fc = the fraction of civilizations that develop a technology that releases detectable signs of their existence into space; L = the length of time for which such civilizations release detectable signals into space. A critical parameter in this equation is "L", the longevity of technological societies measured from the time radio telescopes are invented in an attempt to communicate with other planets. Estimates of "L" range between a minimum of 70 years and 10,000 years, but even for the more optimistic longevity scenario only 2.3 such planets would exist in the galaxy at the present time. Carl Sagan (1980) estimated L on the scale of only a couple of hundred years since a civilization discovered nuclear fission. It is another question whether an organism exists, in this, or any other galaxy, which has triggered a mass extinction of species.

The polarity between gradual biological evolution and mass extinction events parallels the polarity between uniformitarian views of terrestrial history (James Hutton: 1726–1797; Charles Lyell: 1797–1875) and the notion of catastrophism (Cuvier: 1769–1832). Following initial accretion of asteroid and cometary fragments and dust, including amino acid components (Chyba 1993; Chyba and Sagan 1996; Delsemme 2000) and oxygen ($^{18}O/^{16}O$) isotopic evidence from 4.4 Ga zircons suggests that granitic crust formed at that stage was in part cool enough to allow liquid water near the surface (Wilde et al. 2001; Peck et al. 2001; Mojzsis et al. 2001). Cometary seeding of planetary atmospheres is capable of contributing extraterrestrial organic components, incinerated upon impact, possibly leading to shock synthesis of new organic molecules (Chyba and Sagan 1996). Cometary components of terrestrial sediments include aminoisobutayric acid (AIB), isovaline (Zhao and Bada 1989; Zahnle and Grinspoon 1990) and noble gases such as ^{3}H (Farley et al. 1998).

The transformation from organic molecules (amino acids, purine, pyrimidine), to complex information-rich biomolecules (peptide, nucleic acid, protein, enzyme) whose genetic information cannot be expressed by mathematical algorithms, has been estimated as a chance probability of $1:10^{120}$ (Davies 1998). Intrinsic to the question of the origin of early biomolecules is the nature of environmental settings of prebiotic molecules and early microorganisms. Earliest replicating cells at submarine hot springs probably required only twenty or so elements available and as many fundamental organic molecular components (Wald 1964; Eck and Dayhoff 1968). Original biomolecules could have been synthesized from amino acid of both terrestrial and cometary derivation. Panspermia theories, rather than offering an explanation for the origin of biomolecules, only defer the question further back in time and space, reflecting popular fads that advocate extraterrestrial origins of life.

Definitions of life in terms such as "matter that includes responsiveness, growth, metabolism, energy transformation, and reproduction" (Encyclopedia Britannica) and distinctions between animate and inanimate matter are complicated by the existence of intermediate entities, including DNA-free viruses and sub-micron nanobes (Uwins 1998) (Fig. Ep-1). Views of these entities vary, from *"organisms at the edge of life"* (Rybicki 1990) to fragments of DNA and larger cells. Once primordial biomolecules formed, natural selection accounts for their evolution trajectories in terms of mutations, adaptations and self-repair, all the way from microbes to brains. According to Ellis (2005) *"Ever higher levels of interaction and causality arose as complexity spontaneously increased in the expanding Universe, allowing life to emerge. Darwinian processes of selection guided the physical development of living systems, including the human brain"*. However, properties such as directionality and intentionality inherent in evolutionary chains, and the transfer of intelligence, constitute outstanding issues. The question arises, does intelligence constitute the property of organisms and species or, alternatively, does intelligence reside in unknown laws of nature, projected on all life forms. Such a rhetorical question would be in some respects analogous to a question such as "is gravity an inherent property of organisms or a feature inherent in the basic laws of nature?"

Inherent in the question are little-understood top-to-base causality processes (Ellis 2005), directionality and intentionality. Teilhard de Chardin hinted at the existence of laws of complexity giving rise to awareness and consciousness, stating: "... *the higher the degree of complexity in a living creature, the higher its consciousness, and vice versa. The two properties vary in parallel and simultaneously*" (1959, p. 111). The semi-autonomous existence of different

Epilogue: The 'Life Force'

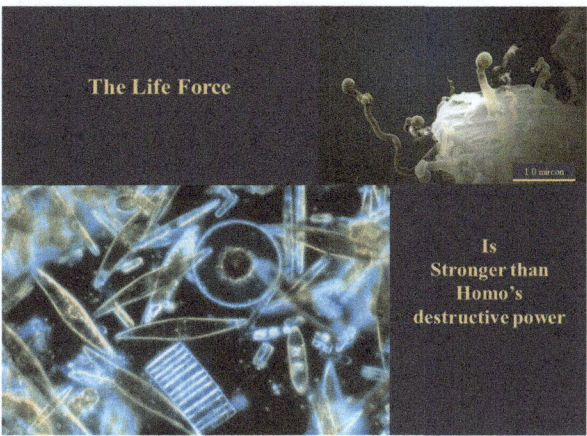

Fig. Ep-1 An epitaph: the life force is stronger than H. sapiens' power. **a** Nanobes, found living in deep fractures (courtesy of P.J.R. Uwins); **b** Diatoms seen through a microscope, encased within a silicate cell wall and living between crystals of annual sea ice in McMurdo Sound, Antarctica (NOAA Corps2365 Collection. Author. G. T. Taylor, Stony Brook University. Source: corp2365, NOAA Corps Collection; http://commons.wikimedia.org/wiki/File:Diatoms_through_the_microscope.jpg

organizational levels within complex systems, including sub-atomic particles, atoms, molecules, biomolecules, nerves and brains, led Ellis (2005) to state: "although the laws of physics explain much of the world around us, we still do not have a realistic description of causality in truly complex hierarchical structures".

Prior to the mastery of fire little difference existed between the behavior of Hominins and intelligent primate species. The mastery of fire and the expression by Homo through burial, art and eventually science, distinguishing the genus from all other creatures, remain to be explained. Acting as a mirror of the world around it, a product of millions of years of evolution, the apparent ability of the human brain to perceive the physical laws and their underlying mathematical logic requires that, embedded in the brain are the very codes it can perceive around it. Where science is based on empirical observations and mathematical calculations, the origin of intuitive and perceptive ideas remains unknown. It may be instructive to examine aspects of human behavior which appear to correlate with physical wave patterns, even if only as metaphors. In quantum mechanics, whereas the statistical behavior of collections of particles is defined by thermodynamic laws, it is not possible to predict the behavior of any single photon or quanta. By analogy, whereas the behavior of populations may follow statistical patterns, the behavior of individuals may be less or even unpredictable, giving rise to the concept of "free will". If so, free will may display an analogy to solitons (solitary waves) formed through interference and amplification of wave patterns, propagating a powerful pulse.

The transition from organisms controlled purely by genetic and instinctive factors to organisms which develop thought processes and cultural traits remains undefined. The intelligent coordination of social systems intrinsic to termite nests, bee hives and modern cities where biological evolution is supplemented or superseded by cultural evolution—remains little understood. DNA and paleontological studies, documenting the molecular and physical evolution of species, are rarely if ever capable of elucidating the progress of intelligence — a little-defined faculty. Major questions remain. Two examples follow:

The resolution of the basic blocks of living organisms, a hierarchy ranging from basic atoms (C, O, H, N), to nucleic acids (adenine, cytosine, guanine, thymine, uracil), to DNA and RNA chains, genes, chromosomes, cell nuclei, multicellular organisms, constituting the hardware of life, leaves the question of the software of life open (Noble 2008). An analogy to the enigma would be discovery by an alien of a computer on a beach. Analyzing its components the alien would resolve its various building blocks, motherboard, electronic circuits, transistors, chips, vacuum tubes, gates—but would be unable to identify the mind which has designed the system. Nor has science to date decoded the natural laws that underlie the origin of the phenomenon of life. The evolutionary chain from DNA/RNA molecules, all the way to the brain, suggests an intelligence written into yet un-decoded laws of complexity and life.

Does the architectural evolution of myriad generations of termite nests represent an evolution of the inherent intelligence, of central "brains" of complex termite "cities"? Do animals and insects act by genetic controls, instincts or even by thought? For example, what is the role of thought where a bunch of termites construct a leaf shelter, each ant having its specific role in the process? Is it possible that humans apply an anthropocentric double standard when differentiating between such a group of ants and a group of people building a house? Is intelligence possessed by individual ants or humans, or does it reside in unknown natural laws? Is it possible that, far from being a unique property of any particular species, intelligence resides in un-decoded laws of nature?

In the absence of explanations, concepts such as that of a "*life force*" (Fig. Ep-1) may be invoked with reference to, for example, the survival of extremophile bacteria. On a larger scale, the Gaia Hypothesis, which views Earth as a single organism maintaining a homeostatic balance (Lovelock 1979), offers an attractive allegory combining both known and unknown elements of planetary evolution. These concepts transcend the boundary between science and philosophy, where the human brain remains in a realm of Flatland (Abbott 1884), blind but beginning, even if too late, to perceive dimensions it cannot comprehend.

We may never know.

Appendices

Tables A.1, A.2, D.1.

Table A.1 Geological stage boundaries, large asteroid impact events, large volcanic provinces and percent mass extinction of species (% mass extinctions after Keller, 2005)

Stage boundaries/ epochs	Large asteroid impacts	Large volcanic provinces	Percentage mass extinction of genera (%)
Mid-Miocene Langhian 15.97 Ma	Ries (24 km) 15.1 ± 1.0 Ma	Columbia Plateau Basalt 16.2 ± 1 Ma	6
Eocene–Oligocene boundary 33.9 ± 0.1 Ma	Popigai (100 km) 35.7 ± 0.2 Ma; Chesapeake Bay (85 km) 35.5 ± 0.3 Ma Mount Ashmore: E-O Boundary	Ethiopian Basalts 36.9 ± 0.9 Ma	10
KT boundary 65.5 ± 0.3 Ma	Chicxulub (170 km) 64.98 ± 0.05 Ma Boltysh (25 km) 65.17 ± 0.64 Ma	Deccan Plateau Basalts. 65.5 ± 0.7 Ma (pooled Ar Ages: 65.5 ± 2.5 Ma)	46
Cenomanian–Turonian 93.5 ± 0.8 Ma	Steen River (25 km) 95 ±7 Ma	Madagascar Basalts 94.5 ± 1.2 Ma	17
Aptian (Early Cretaceous) 125–112 Ma	Carlswell (39 km) 115 ± 10 Ma; Tookoonooka (55 km; 125 ± 1 Ma); Talundilly (84 km; 125 ± 1 Ma); Mien (9 km) 121 ± 2.3 Ma; Rotmistrovka (2.7 km) 120 ± 10 Ma	Ontong-Java LIP 120 Ma Kerguelen LIP 120–112.7–108.6 Ma Ramjalal Basalts, 11 7 ± 1	14

(continued)

Table A.1 (continued)

Stage boundaries/ epochs	Large asteroid impacts	Large volcanic provinces	Percentage mass extinction of genera (%)
End-Jurassic 145.5 ± 4 Ma	Morokweng (70 km) 145 ± 0.8 Gosses Bluff (24 km) 142.5 ± 0.8 Ma; Mjolnir (40 km) 143 ± 2.6 Ma	Dykes SW India 144 ± 6 Ma	20
End-Pliensbachian 183 ± 1.5 Ma		Peak Karoo volcanism Start 190 ± 5 Ma; Peaks 193, 178 Ma; Lesotho 182 ± 2 Ma	19
End-Triassic 199.6 ± 0.3 Ma		Central Atlantic Igneous Province: 203 ± 0.7–199 ± 2 Ma Newark Basalts 201 ± 1 Ma	18
Norian/Rhatian 216.5	Manicouagan (100 km) 214 ± 1 Ma; Rochechouart (23 km) 213 ± 8 Ma;		34
Permian–Triassic: 251 ± 0.4 Ma; 251.4 ± 0.3 to 250.7 ± 0.3 Ma	Araguinha (40 km) 252.7 ± 3.8 Ma	Siberian Norilsk 251.7 ± 0.4–251.1 ± 0.3 Ma	80
Late to end Devonian 374–359 Ma	Woodleigh (120 km) 359 ± 4 Ma; Siljan (52 km) 361 ± 1.1 Ma; Alamo breccia (∼100 km) ∼360 Ma; Charlevoix (54 km) 342 ± 15 Ma	Rifting and 364 Ma Pripyat–Dneiper–Donets volcanism	30 58
End-Ordovician 443.7 ± 1.5 Ma	Several small poorly dated impact craters		60
End-Early Cambrian 513 ± 2 Ma	Kalkarindji volcanic Province, northern Australia 507 ± 4 Ma		42

Table A.2 Comparison of mean global temperature rise rates during the Cenozoic, including the K-T impact events (Beerling et al. 2002), the 55.9 Ma PETM hyperthermal event (Zachos et al. 2008), end-Eocene freeze and formation of the Antarctic ice sheet (34–32 Ma) (Zachos et al. 2001), Oligocene (Zachos et al. 2001), Miocene (Kurschner et al. 2008) and end-Pliocene (Zachos et al. 2001; Beerling and Royer 2011) thermal rises, glacial terminations (Hansen et al. 2007) Dansgaard-Oeschger cycles (Ganopolski and Rahmstorf 2002; Jouzel 2007), 8.2 kyr event (Wagner et al. 2002) intra-Holocene events (IPCC 2007) and Anthropocene climate change (IPCC 2007)

Age	Interval	Mean global land and sea temp change (C)	Warming rate (C/year)	CO_2 change (ppm)	CO_2 change rate (ppm/year)	Reference	Proxy methods*
K-T impact 64.98 Ma	Instant to 10,000 years	Short freeze followed by ~+7.5C	~0.00075	~400–2,300	Instantaneous to 0.19 ppm/yr	Beerling et al. (2002)	Gingko stomata
PETM 55.9 Ma	~10,000 years	~+5 – 9C	~0.0005	~1,800–4,000 ppm	~0.22 ppm/yr	Zeebe et al. (2009)	Deep sea carbonate dissolution
Eo-Ol freeze 34.2 – 34.0	~200,000 years	~-5.4C	-0.000027	~1,120–560 ppm in 10 × 10^6 years		Liu et al. (2009); Pollard and DeConto (2005)	TEX86; $\delta^{18}O$ of benthic foraminifera; Boron and alkenones model
End-Oligocene ~24.7	~200,000 years	~+4C	0.00002	500–900 ppm	0.002	Pekar and Christie-blick (2007)	$\delta^{13}C$ data from alkenones
Mid-Miocene 20 – 18 Ma	~200,000 years	~+1.5C	0.000007	~300 – 520 ppm	0.0011	Kurschner et al. (2008)	multiple-species stomatal frequency record
End-pliocene	4 – 3 Ma	~ +1C	0.000001	~250 – 400 ppm	0.00015	Zachos et al. (2001); Beerling and Royer (2011)	Stomata pores; $\delta^{13}C$ plankton
Glacial terminations/ Eemian	11,000 years	+-5C	0.0004	+100 ppm	0.009	Hansen et al. (2007); Petit et al. (1999); EPICA (2004)	Ice cores
Dansgaard-Oeschger- 21 cycles of ~1,500 years each	~75 – 15 kyr	~3.5C	0.01 – 0.2	+20 ppm	0.2	Ganopolski and Rahmstorf (2002); Jouzel (2007)	Greenland ice cores

(continued)

Table A.2 (continued)

Age	Interval	Mean global land and sea temp change (C)	Warming rate (C/year)	CO_2 change (ppm)	CO_2 change rate (ppm/year)	Reference	Proxy methods*
Younger dryas Interglacial stadial	12.9 – 11.7 kyr	~−15C in GISP2 ice core		−7 ppm			Greenland ice cores
8.2 kyr stadial	~100 years	−3.3C in the North Atlantic		−25 ppm in ~300 years	−0.08	Wagner et al. (2002)	Greenland ice cores
Medieval warm period (MWP)	~400 years	**−0.4 – 0.5C**	**−0.001**	5 ppm	~0.012	IPCC-(2007) Chapter 4	Ice cores, tree rings, cave deposits
Little ice age (LIA)	~60 years	**~−0.4C**	**~−0.006**	−5 ppm		IPCC-(2007) Chapter 4	Ice cores, tree rings, cave deposits
Post-1,750	263 years	**+0.9C + 2.3C potential (with no aerosol masking)**	**−0.0034 −0.008**	280–400 ppm	~0.45	IPCC-(2007)	Instrumental
1975–2012	37 years	**+0.6C**	**~0.016**	330 – 394.28 ppm	~1.73	NASA/GISS IPCC-(2007)	Instrumental
March 2012–2013	1 year			**2.89 ppm**	2.89	NOAA (2013)	

Table D.1 Proposed solar mitigation and atmospheric carbon sequestration methods

Method	Supposed advantages	Problems
SO_2 injections	Cheap and rapid application	Short multi-year atmospheric residence time; ocean acidification; retardation of precipitation and of monsoons
Space sunshades/minors	Rapid application. No direct effect on ocean chemistry	Limited space residence time. Uncertain positioning in space. Does not mitigate ongoing ocean acidification by carbon emissions
Ocean iron filings fertilization enhancing phytoplankton	CO_2 sequestration	No evidence that dead phytoplankton would not release CO_2 back to the ocean surface
Ocean pipe system for vertical circulation of cold water to enhance CO_2 sequestration	CO_2 sequestration	No evidence the cold water would not re-warm when pumped to the surface
"Sodium trees" NaOH liquid in pipe system sequestering CO_2 to Na_2CO_3, separation and burial of CO_2	CO_2 sequestration, estimated by Hansen et al. (2008) at $\sim$ \$300 per ton CO_2	Unproven efficiency: need for CO_2 burial: \$trillions expense (though no more than current military expenses)
Soil carbon burial/biochar	Effective means of controlling the carbon cycle (plants + soil exchange more than 100 GtC/year with the atmosphere)	Requires a huge international effort by a workforce of millions of farmers
serpentine CO_2 sequestration	CO_2 sequestration	Possible scale unknown

About the Author

Andrew Glikson, an Earth and paleo-climate scientist, studied geology at the University of Jerusalem and graduated at the University of Western Australia in 1968. He conducted geological surveys of the oldest geological formations in Australia, South Africa, India and Canada, studied large asteroid impacts, including effects on the atmosphere and oceans mass extinction of species. Since 2005 he studied the relations between climate and human evolution. He was active in communicating nuclear issues and climate change evidence to the public and parliamentarians through papers, lectures and conferences.

References

Abbott E (1884) Flatland: a romance of many dimensions. http://en.wikipedia.org/wiki/Flatland
Beerling DJ, Royer D (2011) Convergent Cenozoic CO2 history. Nat Geosci 4:418–420
Beerling DJ, Lomax BH, Royer DL, Upchurch GR, Kump LR (2002) An atmospheric pCO_2 reconstruction across the Cretaceous-Tertiary boundary from leaf mega fossils. Proc Nat Acad Sci 99:7836–7840
Chyba CF (1993) The violent environment of the origin of life: progress and uncertainties. Geochim et Cosmochim Acta 57:3351–3358
Chyba CF, Sagan C (1996) Comets as the source of prebiotic organic molecules for the early Earth. In: Thomas PJ, Chyba CF, McKay CP (eds) Comets and the origin and evolution of life. Springer Verlag, New York, p 147–174
Davies P (1998) The fifth miracle. Penguin Books, Baltimore
Delsemme AH (2000) Cometary origin of the biosphereē1999 Kuiper prize lecture. Icarus 146:313–325
Drake F, Dava S (1992) Is anyone out there? The Scientific Search for Extraterrestrial Intelligence. Delacorte Print, New York, p 55–62
Eck RV, Dayhoff MO (1968) Evolution of the structure of ferredoxin based on living relics of primitive amino acid sequences. Science 152:363–366
Ellis G (2005) Physics, complexity and causality. Nature 435:743
EPICA Community Members (2004) Eight glacial cycles from an Antarctic ice core. Nature 429:623–628
Farley KA, Montanari A, Shoemaker EM, Shoemaker CS (1998) Geochemical evidence for a comet shower in the late Eocene. Science 280:1250–1253
Ganopolski A, Rahmstorf S (2002) Abrupt glacial climate changes due to stochastic resonance. Physics Rev Lett 88:3–6
Hansen J, Sato M, Kharecha P, Lea DW, Siddall M (2007) Climate change and trace gases. Phil Trans Roy Soc 365A:1925–1954
Hansen J, Sato M, Kharecha P, Beerling D, Masson-Delmotte V, Pagani M, Raymo M, Royer DL, Zachos JC (2008) Target Atmospheric CO2: Where Should Humanity Aim? Open Atmos Sci J 2:217-231
IPCC (2007) Contribution of working group i to the fourth assessment report of the intergovernmental panel on climate change. http://www.ipcc.ch/publications_and_data/ar4/wg1/en/contents.html
Jouzel J et al (2007) Orbital and millennial Antarctic climate variability over the past 800,000 years. Science 317:793–797
Kurschner WM et al (2008) The impact of Miocene atmospheric carbon dioxide fluctuations on climate and the evolution of terrestrial ecosystems. Proc Nat Acad Sci 105:449–453
Liu Z et al (2009) Global cooling during the eocene-oligocene climate transition. Science 323:1187–1190
Lovelock JE (1979) Gaia: A new look at life on earth. Oxford University Press

Mojzsis SJ, Harrison TM, Pidgeon RT (2001) Oxygen-isotope evidence from ancient zircons for liquid water at the Earth's surface 4,300 Myr ago. Nature 409:178–181

NOAA (2013) Mouna Loa CO2. http://www.esrl.noaa.gov/gmd/ccgg/trends/

Noble D (2008) The music of life: biology beyond genes. Oxford University Press, Oxford, p 153

Peck WH, Valley JW, Wilde SA, Graham CM (2001) Oxygen isotope ratios and rare earth elements in 3.3 to 4.4 Ga zircons: Ion microprobe evidence for high $\delta^{18}O$ continental crust and oceans in the Early Archaean. Geochim et Cosmochim Acta 65:4215–4229

Pekar S, Christie-Blick N (2007) Showing a strong link between climatic and pCO_2 changes: resolving discrepancies between oceanographic and Antarctic climate records for the Oligocene and early Miocene (34 − 16 Ma). Univ. Nebraska Antarctic Drilling Program. http://digitalcommons.unl.edu/andrillaffiliates/17

Petit JR et al (1999) 420,000 years of climate and atmospheric history revealed by the Vostok deep Antarctic ice core. Nature 399:429–436

Pollard D, DeConto RM (2005) Hysteresis in Cenozoic Antarctic ice sheet variations. Glob Planet Change 45:9–21

Rybicki EP (1990) The classification of organisms at the edge of life, or problems with virus systematics. S Afr J Sci 86:182–186

Sagan C (1980) Cosmos. Macdonald Futura Publishers, London, p 365

Uwins PJR et al (1998) Novel nano-organisms from Australian sandstones. Am Mineral 402 83:1541–1550

Wagner F, Aaby B, Visscher H (2002) Rapid atmospheric CO2 changes associated with the 8,200-years-B.P. cooling event. Proc Nat Acad Sci 99:12011–12014

Wald G (1964) The origin of life. Proc USA National Acad Sci USA 52:595–611

Wilde SA, Valley JW, Peck WH, Graham CM (2001) Evidence from detrital zircons for the existence of continental crust and oceans on the Earth 4.4 Gyr ago. Nature 409:175–178

Zachos J, Pagani M, Sloan L, Thomas E, Billups K (2001) Trends, rhythms, and aberrations in global climate 65 Ma to present. Science 292:686–693

Zachos J, Dickens GR, Zeebe RE (2008) An early Cenozoic perspective on greenhouse warming and carbon-cycle dynamics. Nature 451:279–283

Zahnle K, Grinspoon D (1990) Comet dust as a source of amino acids at the cretaceous/tertiary boundary. Nature 348:157–160

Zeebe RE, Bada JLZachos JC, Dickens GR (2009) Carbon dioxide forcing alone insufficient to explain palaeocene–eocene thermal maximum warming. Nat Geosci 2:576–580

Zhao M, Bada JL (1989) Extraterrestrial amino acids in cretaceous/tertiary boundary sediments at Stevns Klint, Denmark. Nature 339:443–445

Index

A
Aboriginal 'fires-tick farming', 82
Acraman impact and Acritarchs radiation, 49, 50
Acraman/Bunyeroo, 48
Acritarchs
    radiation of, and Acraman impact, 49, 50
    stratigraphic distribution of, 50f
Ammonia-releasing microbes, 8
Archaean atmosphere and hydrosphere, 7
Asteroid impact events of species, 155, 156t
Atmosphere
    air-ocean system, chemical changes in, 3
    Archaean, see Archaean atmosphere and hydrosphere
    biological activity, 3
    CO_2, 48
    Cenozoic atmospheres, see Cenozoic atmospheres and early hominins
    during early history Earth, 6
    fire, 71
    great carbon oxidation event, see great carbon oxidation event
    late Proterozoic, 14, 15
    Palaeozoic and Mesozoic, see Palaeozoic and Mesozoic atmospheres
Atmospheric carbon
    CO_2 levels, 72
    carbon sequestration methods, proposed solar mitigation and, 160t
Australopithecines garhi, 75
Australopithecus africanus, skeleton of, 80f

B
Banded iron formations (BIF), 7, 10, 11f
    sub-glacial anoxia, 14
    transportation of ferrous iron, 8
Baragwanathia longifolia, 72f

Basaltic crust, 6
Bergeroniellus asiaticus, 16f
Bethe, Hans, 150
Bikini H-bomb tests, 150
Bush fires, threat of, 147

C
Cainozoic CO_2
    multi-proxy-based trends, 32f
    and temperature trends, 31f
Cambrian
    arthropods, 16f
    explosion, 15
    occurrence of ice ages and cold phases, 22f
Canfield Ocean, 15
Carbon and oxygen isotopes and mass extinctions
    Ashgill/Hirnantian glaciation and extinction event, 60
    Central Atlantic Magmatic Province (CAMP), 63
    chemostratigraphic profiles, 64f
    $\delta^{13}C$ values, 60
    comparison, Canning Basin and Europe/Canada, 62f
    across K-T boundary, 65
    in Montana and Wyoming, 65
    general carbon isotope trend, 63f
    isotopic changes in late Ordovician, 61f
    oxygen isotope ratios, 60, 61
    prolonged warming, impact on, 63
    Siberian Trap volcanism, 61
    total organic carbon (TOC) and stable carbon isotope ratios, 59
    upper Devonian isotopic excursions, 60
Carbon cycle, schematic illustration of, 23, 24f

Carbon dioxide (CO_2), concentrations and radiative forcing by, 96f
Carboniferous-Permian period, 73
　glaciations, 23, 25, 26
　interval, 21
Cenozoic, 21, 23, 25
　mean global temperature rise rates, 158, 159t
Cenozoic atmospheres and early hominins
　Antarctic ice sheet formation
　　and global climate, 30
　　between 34 and 33.5 Ma, 30
　Cainozoic CO_2
　　multi-proxy-based trends, 32f
　　and temperature trends, 31f
　climate and hominin evolution, 41f
　East African rift valleys, 42f
　Eocene cooling, 32
　　end-Eocene freeze, 34f
　glacial-interglacial cycles, 39f
　hominins, evolution of, 37, 38
　human evolution and climate transitions, 41, 42
　Intertropical Convergence Zone (ITCZ), 37, 38
　intra-glacial Dansgaard-Oeschger (D-O) cycles, 40f
　late Miocene (9.3-8.4 Ma) sediments, 38f
　Milankovic cycles and Cenozoic paleogeography, orbital components of, 35f
　North Atlantic Thermohaline Current (NATH), 40
　Paleocene-Eocene thermal maximum (PETM), 29, 33f
　Pliocene and late Holocene climates, 36, 37f
　sea surface temperatures, evolution of, 39f
Charcoal, 71, 72
Classic Maya civilization, incipient collapse of, 99f
Climate
　changes, 83, 124, 125
　　and greenhouse gases, 121
　　and habitats, 128
　　during Pleistocene, 79
　　disruptions, with last glacial termination, 92
　fire, impact on, 81
　global warming and ocean acidification, see global warming and ocean acidification
　and Holocene civilizations, see Holocene civilizations and climate
　early Anthropocene hypothesis, 97f
　and hominin evolution, 41f, 77f

　models, 57
　　based on mid-Cenozoic conditions, 25
　sea level rise and, 113
　shift, 41, 118
　transition, 41, 77, 78
　variability, 38, 41, 77, 78
　warming, 12
Collider, Hadron, 151
Compton, Arthur, 150
Cosmic collision, 5
Cosmic radiation, 5, 6
Cradle of Mesopotamian, 95
Cretaceous-Tertiary boundary, 49
　K–T mass extinction, 56, 57
Cryogenian Snowball Earth, 13
　evidence of, 14f
　cap carbonate, 14
　Paleomagnetic evidence, 14
　negative carbon isotope anomalies, 14
　late Proterozoic, 14
Cycads, 71

D

Deep hot biosphere, 6
$\delta^{18}O$ value, 7
Dickinsonia costata, 15f
Dresser Formation, 10, 13f
Dunbar index, 76
　see also neocortex

E

Early Anthropocene hypothesis, 97f
Early Jurassic, 49, 54
　to Cretaceous cool phases, 26
　Pliensbachian, 49
Early Mississippian, 72
Early Proterozoic Huronian glaciations, 13
Earth-Mars-Venus comparison, 4f
East African rift valleys, 42f, 77
Eccentricity, 32, 35f, 37, 38
　variations of, 41f, 146f
Ediacara fossils, 15f
Ediacaran complex Acritarchs palynoflora (ECAP), 49, 50f
Ediacaran leiosphere palynoflora (ELP), 49, 50f
Emerging bipolarity, 85f
End-Devonian, 49
　mass extinctions, 51, 60
End-Eocene freeze, 31f, 34f, 59, 158t
End-Jurassic, 49, 156t
End-Triassic, 25, 49, 156t

Index

mass extinction, 54, 63
Eoandromeda octobrachiata, 15f
Eocene cooling, 32
Evaporite deposits, 7
Evolution
    of biodiversity over geological timescales, 73
    of CO_2 and occurrence of ice ages and cold phases, 22f
    of hominins, 76
        in rift valley region of Eastern Africa, 37
    of human brain, 78f
    of life through the Phanerozoic era, 23f
    of marine invertebrate, mass extinction events, 47
    of oxygen, stages in, 24, 25
    of sea surface temperatures, 39f

F

Faint Early Sun conditions, 13
Fear, 84, 85, 151
Flooding of low-lying river systems, 146
Formation
    of Antarctic ice sheet, 158t
        and global climate, 30
        between 34 and 33.5 Ma, 30
    of BIFs, 10
    of lakes and sheltered environments, 37
    of metallic core, 5
    of sedimentary iron, 14
Fossil bovids, Africa-wide occurrences of, 76f
Fossil palaeozoic plants, 72f
'founder effects', 88

G

GEOCARBSULF model, 24
Geological stage boundaries of species, 155, 156t
Geological time scale, evolution of, 5
Ginkgo, 71
Glacial-interglacial cycles, 29, 37, 39f, 137
Global temperature, 30, 32, 115, 125f
    and climate forcings, 92f
    historical and projected variations in, 139, 140f
    rise in, 40, 111
        during Cenozoic, 158, 159t
    sulphur aerosols, 110
    sulphur emissions, effects on, 112f
Global warming and ocean acidification, 133
Atmosphere–Ocean Global Circulation Model (AOGCM) projections, 139, 140f
atmospheric CO2 concentrations, 134f
carbon dioxide and global mean climate system changes, 139, 140f
CO2 draw-down methods, 142
    principal observations, 142, 143
CO2 emissions, 134f
    estimates of fossil fuel resources and equivalent atmospheric CO2 levels, 135f
    factors inducing, 133
        carbon stored in Arctic and boreal regions, 134, 135
        changes in ENSO and Pleistocene glacial-interglacial cycles, 137
        methane deposits from permafrost and frozen Arctic lakes and Sediments, 135, 136
        North Atlantic Thermohaline Current (NATH), 136
        reserves of fossils fuels, 133, 134
global distribution of methane hydrate deposits, 137f
historical and projected variations in Earth's surface temperature, 139, 140f
methane bubbling, 137f
ocean circulation strength, change in, 138f
potential policy-relevant tipping elements, 139, 140f
pre-Oligocene (pre-34 Ma) ice-free conditions, 142
21st century climate trends, 142
vulnerable carbon sinks, 136f
Granite-dominated crustal nuclei, 6
Great carbon oxidation event, 107–126
    albedo-flip feedback process, 111
    atmospheric CO2, methane, nitrous oxide, trends in, 109f
    Australian CSIRO report, 110
    carbon at Paleocene-Eocene boundary, 124
    CO2 rise rates and mean global temperature rise rates, 115f
    carbonic acid (H_2CO_3), 119
    Cenozoic climate shifts, 111
    chlorophyll in oceans, 122f
    climate effects and GHG, 121
    current Earth radiative imbalance, 116
    early Eocene condition versus existing condition, 111
    extreme weather events, 119f
    fractionation of CO_2 emissions, 119f

global carbon budget 2002-2011, 110f
Global Carbon Project, 107
global CO_2 emission estimate, 107f
global mean radiative forcings, 108f
global warming, 115f
    ENSO cycle, 125f
    positive feedbacks, 122
Greenland mass loss, 113
ice melt in Greenland and Antarctic, 116f
La-Nina cycles, 122
local temperature anomalies, 121f
monthly mean global surface temperatures
ocean acidification, 123, 124f
ocean heat contents, 115f
relations between sea levels and temperatures, 117f
sea level rise, 113
    of historic and 20th to 21st centuries, 118f
sea water temperature rise, 113
and sulphur emissions, 112f
tipping point, 118
United States, 126
winter precipitation trends in Mediterranean region, 120f
WWI and WWII, emissions of CO2 and SO2, 107
Greenhouse gases (GHG), 3, 48, 57, 101f, 106, 107, 108, 109f, 134, 135, 142
    atmospheric levels, 8f, 13, 41, 47
    mid-Holocene rise in, 91
    rise in, 111, 143
    temporal fluctuations in, 6
Greenland Summit, surface temperature response at, 94f

H
Hadean era, 5
High-temperature conditions, 7, 71
Holocene civilizations and climate
    Classic Maya civilization, incipient collapse of, 99f
    climate disruptions, with last glacial termination, 92
    cradle of Mesopotamian, 95
    early Anthropocene hypothesis, 97f
    global temperature and climate forcings, 92f
    Greenland Summit, surface temperature response at, 94f
    last glacial termination, d (deuterium) ‰ variations, 93f
    late Holocene climate perturbations, 95
    medieval drought in Southeast Asia, 100f
    Medieval Warm Period (MWP), 98
    Nile River, stages in history of, 93, 94
    Older dryas, 91
    Quelccaya ice cores, 97
    radiative forcings
        CO_2, CH_4 and N_2O concentrations, 96f
        during last 1.1 kyr, 101f
    rise of ancient river civilizations, Egypt and Babylon, 98f
    Sahara peak Holocene rock painting, 95f
    simulated temperatures, during last 1.1 kyr, 101f
    Younger dryas, 91
Hominid, 29
Hominines, 75
Hominins, 73f, 78, 81, 82
    in Africa and subsequent migrations, 75
    ancestors of modern humans, 42f
    behavior of, and intelligent primate species, 161
    climate and, 77f
    evolution of, 29, 76
        and climate, 41f
        in rift valley region of Eastern Africa, 37, 38
Homo
    bipedalism, 75
    brain mass, 75
Homo erectus (H. *erectus*), 38, 75, 80, 81, 82, 82t
Homo ergaster (H. *ergaster*), 38, 80f, 81, 82t
Homo floresiensis, 88
Homo habilis (H. *habilis*), 38, 75, 80f, 82
Homo heidelbergensis (H. *heidelbergensis*), 75, 80, 80f
Homo neanderthalensis (H. *neanderthalensis*), 75, 80, 80f
H. *rudolfensis*, 38
Homo sapiens, 79, 86
    oxygen-carbon cycle, 150
    cranium of, 80f
Human brain, 75, 149, 161, 162
    evolution of, 78f
    human brain/body mass ratio, 76
Human evolution and climate transitions, 41, 42, 77
Human migration model, based on mitochondrial DNA, 83f
Human use of fire, prehistoric sites containing evidence, 82t

Index

I
Intertropical Convergence Zone (ITCZ), 37, 38, 78
Intra-glacial Dansgaard-Oeschger (D-O) cycles, 40f

J
Jurassic, 6, 26
   to late Cretaceous, CO_2 and temperature records for, 56f
Jurassic-Cretaceous extinction, 54, 56

K
K-T (Cretaceous-Tertiary boundary) mass extinction, 56, 57
   at El Kef, Tunisia, 58f
1990 Kyoto Protocol, 84

L
Large volcanic provinces of species, 155, 156t
Last glacial termination (LGT), 83, 92
   d (deuterium) ‰ variations, 93f
Late and end Devonian, 156t
   to early Carboniferous glaciations, 23
   mass extinctions, 51
      biotic effects of, 52f
Late Devonian (Frasnian-Fammenian), 48
Late Devonian-early Carboniferous glaciations, 25
Late Heavy Bombardment (LHB) on the Moon, 5
Late Holocene climate perturbations, 95
Late Miocene (9.3-8.4 Ma) sediments, 38f
Late Ordovician
   glaciations, 23, 25
   mass extinction, 50
      CO_2 and temperature records for, 53f
Late Permian and Permian–Triassic mass extinctions, 51–54
   major extinction phases, 51, 52
Late Proterozoic Cryogenian glaciations, 13
Late Silurian, emergence of land plants in, 71
Late Triassic, 49
Leanchoilia superlata, 16f
Life force
   concept of, 162
   and damaging power, 152f
Living planet index 1970-2000, 144, 145f
Long interglacial period, 145, 146f

M
Maori colonization, 83
Marella splendens, 16f
Mass extinction species, 47–49, 143, 150, 155, 156t
   Acraman impact and Acritarchs radiation, 49, 50
   carbon and oxygen isotopes and mass extinctions, *see* carbon and oxygen isotopes and mass extinctions
   end-Eocene freeze, 59
   end-Triassic mass extinction, 54
   Jurassic-Cretaceous extinction, 54, 56
   K-T mass extinction, 56, 57
   late and end Devonian mass extinction, 51
   late Ordovician mass extinction, 50
   late Permian and Permian–Triassic, *see* late Permian and Permian–Triassic mass extinctions
   Paleocene-Eocene extinction, 57–59
   Phanerozoic mass extinction, 48f
   sixth mass extinction, *see* Sixth mass extinction of species
Mass-independent fractionation of sulphur isotopes (MIF-S), 7
Maya blood cult, 86
Medieval drought in Southeast Asia, 100f
Medieval Warm Period (MWP), 98
Methane (CH_4), 6, 7, 14, 29, 32, 47, 48, 57, 59, 60, 107, 120, 122, 141–143
   atmospheric trend, 109f
   concentrations and radiative forcing by, 96f
   deposits, 135
   eruptions, 3
   global average forcings, 108f
   photolysis of, 7
   rise, 140
Methane clathrate deposits, 71
Methane hydrate deposits, 71, 136f, 139, 140f
   global distribution of, 137f
Microbial methanogenesis, 7, 8
Microbial sulphur metabolism, 8
Mid-Pleistocene climate transition, 78
Milankovic cycles and Cenozoic paleogeography, orbital components of, 35f
Milky Way galaxy, 157
Mungo man No. 3, 85f
Mutual Assured Destruction (MAD), 151

N
Neocortex, 85, 86
   to brain ratio, 78f, 79f

Neolithic burning and early global warming, 105–107
  North American Indians, 105, 106
  North American fire regimes, 106
  human-dominated Anthropocene era, 106, 107
  CO2 and methane, mid-Holocene rise of, 107
Neolithic human civilization, 79
Nile River, stages in history of, 93, 94
Nitrous oxide (N_2O), concentrations and radiative forcing by
North Atlantic Thermohaline Current (NATH), 40
Northern Hemisphere glaciations, 78

O
Older dryas, 91
Oppenheimer, Robert, 150
Oxygen, 3, 73, 74, 150
  anoxia, 48
  in Archaean atmosphere and hydrosphere, 7
  atmospheric, 135
    and stromatolites, 12, 13f
  capture by iron-oxidizing microbes, 8
  deep-sea isotopes, 34f
  in DNA species, 149
  drop in, 53
  in Earth, early history, 6
  evolution stages, 24, 25
  and fire, 71
  isotopes, see carbon and oxygen isotopes and mass extinctions
  Phanerozoic geochemical models of, 21
  and photosynthesis, 7, 23
    and glaciation, 13
    photosynthesizing colonial organisms, 8
  proteins binding, 15
  variations through Phanerozoic, 23f, 24
Oxygen isotope data, 9f
  see also carbon and oxygen isotopes and mass extinctions
Ozone layer, 7, 8
  depletion, 10, 57
  ozone hole growth, 118
  molecular resonance of GHG, 32, 108f
  UV-induced isotopic fractionation, 10f

P
Pale blue dot, the, 4f
Palaeozoic and Mesozoic atmospheres

carbon cycle, schematic illustration of, 23, 24f
carbon isotope composition of, 21
Carboniferous-Permian
  glaciations, 23, 25, 26
  interval, 21
early Jurassic to Cretaceous cool phases, 26
evolution of
  CO_2 and occurrence of ice ages and cold phases, 22f
  oxygen, stages in, 24, 25
GEOCARBSULF model, 24
late Ordovician glaciations, 23, 25
oxygen variations through phanerozoic, 23f
Phanerozoic
  evolution of life through, 23
  geochemical models, 21
  upper Cenozoic post-32 Ma glaciations, 21
Paleocene-Eocene boundary, 124
Paleocene-Eocene extinction, 57, 59
Paleocene-Eocene thermal maximum (PETM), 29, 33f, 57, 115f
Paranthropus boisei-nairobi, cranium of, 80f
Parvancorina minchami, 15f
Pennsylvanian, 72
Percent mass extinction of species, 155, 156t
Permian–Triassic boundary, 23, 49, 59, 60, 156t
Permian–Triassic crisis, 63
Permian–Triassic mass extinction, 51–54, 53f
  geochemical trends, 52, 53
Phanerozoic
  atmospheric CO2, 22f
  atmospheric history
    Carboniferous-Permian glaciations, 25, 26
    early Jurassic to Cretaceous cool phases, 26
    late Devonian-early Carboniferous glaciations, 25
    late Ordovician glaciations, 25
  evolution of life through, 23
  geochemical models, 21
  mass extinction, 48f
    carbon and oxygen isotopic anomalies, 59
Planetcide, 151
Pliocene and late Holocene climates, 36, 37f
Ptychagnostus atavus, 16f

Q
Quanta against individual behavior, 161, 162
Quaternary and holocene charcoal records, 73f

Index

Quaternary records, 87
  Anthropocene, 146
Quelccaya ice cores, 97, 99f

R
Radiative forcings
  and concentrations, 96f
  global mean positive and negative, 108f
  by greenhouse gases, 109f, 133, 142
  during last 1.1 kyr, 101f
  rise in, 122
Resolution of basic blocks of living organisms, 162
Rise of
  agricultural civilizations, 106
  ancient river civilizations, Egypt and Babylon, 98f
  anthropogenic greenhouse gas levels, 106
  atmospheric CO2 levels, 51, 54, 57, 110, 111, 133, 145
  atmospheric oxygen, 74
  CO2 and methane, 107
  Hominins in Africa, 75
  land plants, 98f
  Panama cordillera, 33
  sea surface temperature, 137

S
Sahara peak Holocene rock painting, 95f
Sahelanthropus tchadensis, 76, 80f
Sea surface temperatures (SST), 36, 37f, 92f
  evolution of, 39f
Sedimentary rock types, carbon $\delta^{13}C$ values 11f, 11f
Self-repairing DNA-RNA biomolecules, 149
SIAL (Silica-Alumina-dominated crust), 6
SIMA (Silica-Magnesium-iron-dominated crust), 6
Simulated temperatures, during last 1.1 kyr, 101f
Sixth mass extinction of species, 126–129
  relative loss of biodiversity of vascular plants, 128f
  conversion of terrestrial biomes, 128f
  threatened vertebrates, 128f
Small human clans, Pleistocene, 79
Species extinction rates, 145f
  see also Sixth mass extinction of species
Sprigina holotype, 15f
Stromatolites, 8, 12, 13f
Sulphur, 71
  aerosols, 108, 110, 142
  emissions on global temperatures, 110, 112f, 121, 142
  isotopic analyses, 7
    fractionation values for, 10f
  isotopic values, 64
  MIF-S, 7
Sulphur cycle, 21
Sulphur-reducing microbes, 54

T
Taung Child, cranium of, 80f
Teller, Edward, 150
Terrestrial climates, 3, 32
Theia, 5
Thin breathable veneer, formation, 3
Toba volcanic eruption, 86–88, 87f
Tools and construct articulate structures, 76
Triassic fauna, 63
Triassic-Jurassic boundary, 55
  palaeo-atmospheric CO2 variations, 55f
  see also end-Triassic; late Triassic; Permian–Triassic boundary; Permian–Triassic crisis; Permian–Triassic mass extinction
Tribrachidium heraldicum, 15f

U
Under A Green Sky, 143
Upper Archaean glaciations, 13
Upper Cenozoic post-32 Ma glaciations, 21
Upper Devonian isotopic excursions, 60
Upper Eocene glaciations, 30

V
Vascular plants, 71
  relative loss of biodiversity of, 127, 128f
Volcanic
  activity, 7, 8, 101f
  aerosols, 87, 108
  ash, 87
  and/or asteroid events and mass extinction, 48, 49
  eruptions, 3, 24, 48, 60, 72, 86, 149
  events, 22f, 83
  outgassing, 14
  provinces, *see* large volcanic provinces of species
  Toba eruption, *see* Toba volcanic eruption
'volcanic winter' effects, 87f, 88

W
Walker Circulation, 37, 78
War of the Worlds (Wells), 151
Wildfires
    fire-resistant plants, 74
    global fire activity through time, 73*f*
    during Paleozoic and Mesozoic, 73
    prior to mastery of fire by Hominins, 72
    in terrestrial biogeochemical cycles, 72, 73
World War I (WWI), 86
World War II (WWII), 86, 143

Y
Younger dryas, 91

The manufacturer's authorised representative in the EU is Springer Nature Customer Service Centre GmbH, Europaplatz 3, 69115 Heidelberg, Germany. If you have any concerns regarding our products, please contact ProductSafety@springernature.com

Printed and bound by CPI Group (UK) Ltd, Croydon, CR0 4YY

25/03/2026

02078223-0004